AF798563

TAKAHĒ

Geoffrey 'Doc' Orbell at the 50th anniversary of the 1948 rediscovery of takahē. **ROD MORRIS**

For Malcolm

Department of Conservation
Te Papa Atawhai

This book was published with the support
of the Department of Conservation's
Takahē Recovery Programme

TAKAHĒ

BIRD OF DREAMS

ALISON BALLANCE

pb potton & burton

CONTENTS

TAKAHĒ SITES

TE IKA-A-MĀUI / NORTH ISLAND

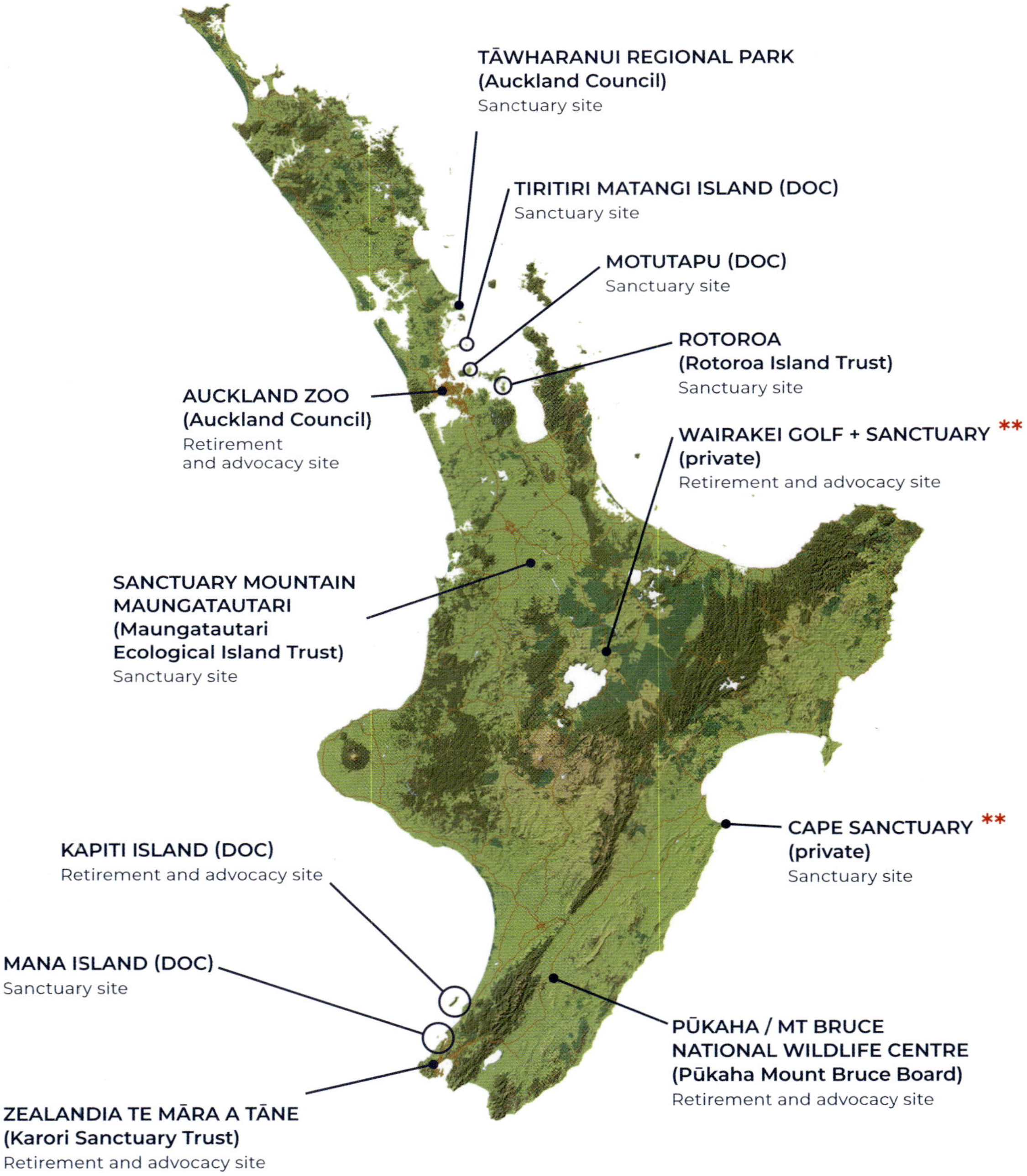

TE WAIPOUNAMU / SOUTH ISLAND

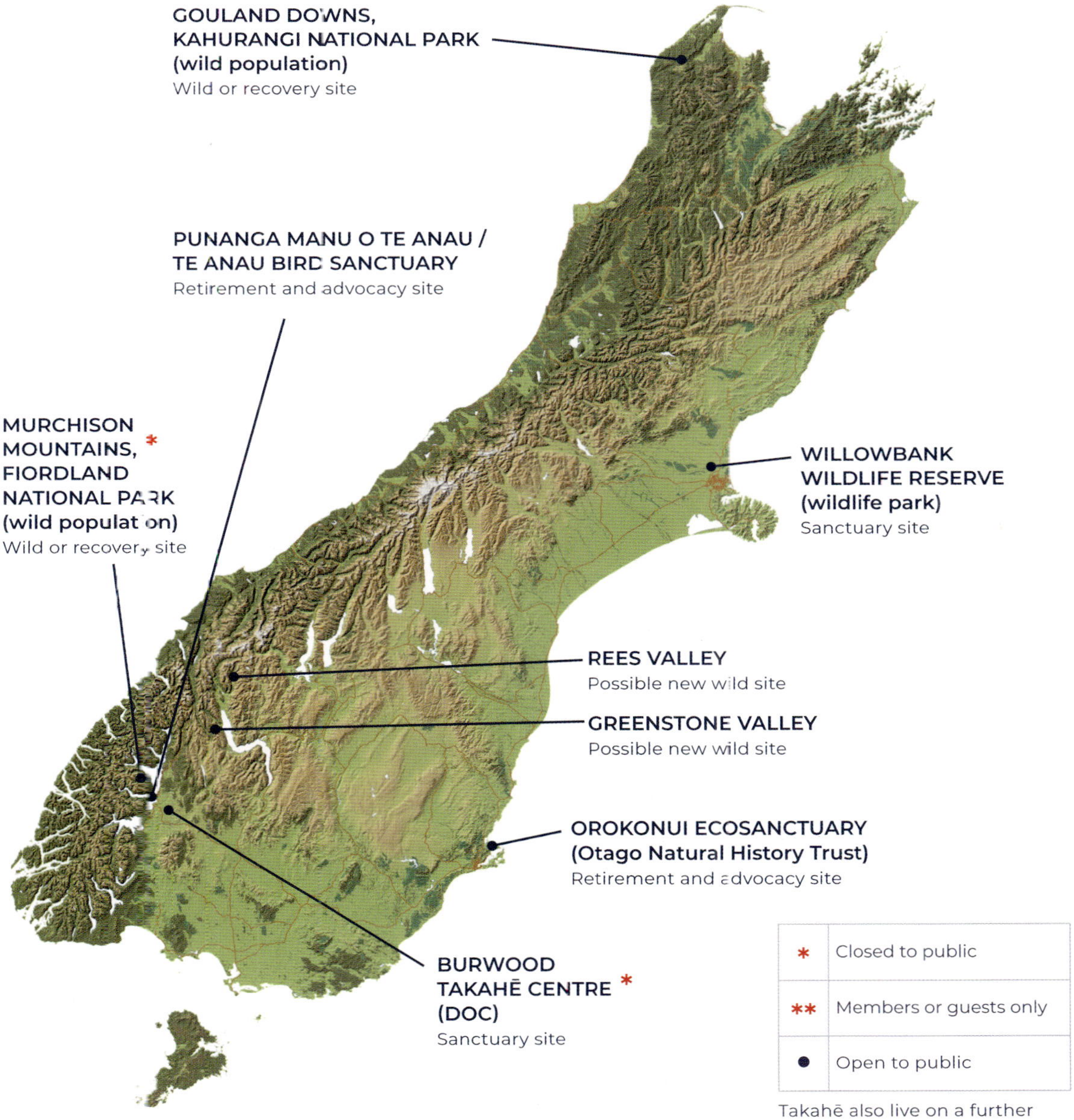

Takahē also live on a further two privately-owned islands.

PREFACE

For several summers in the early 1980s I worked in Fiordland National Park as an interpreter, a story-telling guide, on the busy summer holiday programme. One of our most popular trips, which we offered only once each summer, was to an area that was usually off-limits to the general public. It involved a boat ride across Lake Te Anau to the glow-worm caves jetty, and then a steep climb up the Tunnel Burn to Takahē Valley. We were walking on hallowed ground, retracing the footsteps of those involved in one of New Zealand's legendary bird stories. This is a story that offers everything – loss, rediscovery, hope, disappointment, perseverance and hard-won successes. And there was always a simmering excitement and a sense of anticipation in the crowd – even though we never actually saw a takahē.

Fortunately for us all, there was a new opportunity in the township of Te Anau to see the mysterious bird in real life – at the Te Anau Wildlife Park (now the Punanga Manu o Te Anau / Te Anau Bird Sanctuary). The park was, and remains, a small affair. There are a few aviaries providing homes to birds too injured to be returned to the wild, and a large circular fenced enclosure that is mainly grass with a few areas of bush to provide the residents some privacy if required. At the time, the Takahē Recovery Programme was at a turning point. A long phase of field research was coming to an end and in its place an intensive management programme, which included captive rearing, was just beginning. The pioneering project had several years under its belt and the first results – several hand-reared takahē – were in residence in the pen.

The takahē were a marvel. Described by a leading American ornithologist as resting 'comfortably in the company of elephant birds and moa as one of the world's most strange and wonderful avian beasts', takahē are big – the size of a fat turkey.[1] They are elegantly colourful, a mix of purple-blue, dark blue, turquoise, and emerald and olive greens, with a white rump and a startling red bill and legs. Most New Zealand birds have subtle, even dull, coloration, but not the takahē. The bill and the prominent frontal plate above it looks industrial strength, while the sturdy legs are worthy of a rugby player. It was the stately

A male takahē walks through tussock and *Dracophyllum* subalpine scrub in his Takahē Valley territory. Murchison Mountains. ROD MORRIS

demeanour of the birds in the Te Anau sanctuary that appealed the most, though. They exuded a sense of calm obliviousness to what was happening outside the fence, going about their business in a leisurely, dignified manner that nonetheless had a singular intense focus: continuous feeding. And they were surprisingly chatty, quietly and continuously grunting to stay in contact with one another.

The Takahē Recovery Programme was established alongside two similar schemes to protect the 'big three' threatened bird species in Aotearoa New Zealand at the time, the others being the black robin and kākāpō. What was happening with all three species would determine whether or not they survived. What intrigued me about takahē was the sense of rich evolutionary and human history that went with the bird, paired with the contemporary excitement of world-leading threatened species conservation. I remember visiting the wildlife sanctuary one day, to find Wildlife Service officer Nick Torr standing next to the empty enclosure holding an aerial above his head. He told me he was always being asked by curious park visitors what TV channel he was trying to find. But in the

Dana Bivvy is in the Dana Peaks, sited in a basin off a side branch of the Ettrick Burn. A second small portable bivvy acts as entrance way and storage room. PHIL MARSH

public's defence, it was the very early days of radio-tracking and no one knew that he was locating birds from signals given out by a small item of cutting-edge electronic wizardry strapped to their back. As I watched, he received the reassuring beeps confirming that all the takahē were in residence, just temporarily out of sight.

The Te Anau takahē were a popular tourist attraction, just as the birds were at the only other site they were on display, the Mt Bruce Native Bird Reserve (now the Pukaha National Wildlife Centre) in the Wairarapa. However, the main focus for takahē conservation was not the handful of display birds, but the original wild population in its ancestral seat in the Murchison Mountains. In late 1984, I was delighted when I was allowed to join the newly formed Wildlife Service takahē team on one of their regular spring surveys in the 'Murchies', as they are commonly known. We worked in the subalpine zone, on high ridges and head basins, as well as down in the valleys – basically, we searched everywhere except the forest and scree slopes. The country was steep, rugged and spectacular, the vegetation was thick and scratchy, the weather was not always kind, and our shelters for the night were a succession of small – very small – huts.

The aim was to check on every pair of 'tarks', to find out if they were nesting and whether any eggs were fertile. The team knew each takahē territory and, working in pairs for safety, it was a case of visiting them in turn and looking for sign. Takahē don't try to cover their tracks – they leave a swathe of felled tussock tillers in their wake and a generous trail of droppings. I soon learnt to differentiate fresh green sign from old dried sign that can stick around for several years, and discovered that a mound of droppings indicated the latrine of a nesting bird. Who knew that a toilet was such an important clue! I was more hindrance than help, and was constantly impressed by Nick's ability to find the one tussock bush amid thousands that sheltered a nest. Finally, I was the one to find a latrine and I stooped to part the closest curtain of arching tussock stems. To my surprise (I wasn't expecting to find it straight away), I found myself face to face with a nesting takahē. We stared at each other for a few long seconds, and its frank gaze reminded me that here was a bird that was a survivor, belonging to a species that had twice come back from the grave – but one whose continued existence was hanging in the balance.

I met that takahē almost 36 years to the day since the species was famously rediscovered in the Murchison Mountains by Geoffrey Orbell in 1948 after twice being declared extinct. It was a key time, nearly halfway between rediscovery and the present day.

A takahē just visible on a nest beneath a large tussock. To remain incognito, the incubating bird tucks its red beak out of view. Incubation is early on as the latrine, marked by white droppings in the front of shot, is still small. ANDREW DIGBY

There were just 142 birds in the Murchisons in 1984, and a mere handful more in captivity. It had been – and would continue to be – a long, slow road to recovery for the rare bird. Long-time takahē ranger Dave Crouchley told me that when he started working with takahē in the early 1980s, he was told by a colleague that this was a species that experienced 'periodic plummets'. For decades, the takahē population seemed to stubbornly sit between 100 and 200 birds, and those plummets made numbers spiral down as often as they climbed.

It is now 75 years since the rediscovery, and a lot has happened since then to set the takahē up for a much brighter future. The Takahē Recovery Programme is the longest-running species conservation programme in New Zealand, and perhaps even in the world. It has bravely trialled new approaches and led the way in many aspects of threatened species conservation. Not everything worked as planned, but lessons were learnt and progress made. The initial focus was on the Murchison Mountains, and while the wild takahē population is still the jewel in the crown, the recovery programme these days is more like a three-legged stool. The Fiordland population is the wild heart, but equally important are the captive breeding programme at the Burwood Takahē Centre, and the insurance populations on islands and in sanctuaries around the country. The recovery programme would like to add extra legs for future stability, and that's where new wild sites – known as recovery sites – come into the picture. The takahē diaspora has taken the birds far beyond their natural range, into new and often highly modified habitats. This sets these translocations apart from those of most other New Zealand threatened bird species, which are generally moved from one forest habitat to another similar forest habitat.

The takahē team and I were hoping that the takahē population would pass the magic mark of 500 birds before this book came out. It got very close: 475 birds. But by the time the 75th anniversary is marked, in late 2023, there will definitely be more than 500 birds, a milestone that deserves a real celebration. It finally feels as if the many dedicated researchers and wildlife rangers who have contributed over the years are getting close to seeing the goal they set for the species come to pass: that takahē exist in growing numbers in large areas of their former natural range as a functioning element of natural ecosystems, and are treasured as a national icon.

How we've got to this point is what this book is about. It charts the history of the Takahē Recovery Programme, investigates the species' place in the bird world and reminds us how close takahē came to extinction. But let's begin by pulling together everything we've learnt about the biology and lifestyle of this delightful bird, which can be summed up as a food-obsessed, serially monogamous (mostly), devoted partner and parent, whose teenage kids stick around to help raise their younger siblings. And while these 'big blue chickens' might not be the brightest birds in the flock, they excel at being takahē – and our lives are richer for having them around.

With its red beak and prominent frontal shield, strong red legs, and feathers that range in colour from blue, purple and green, the takahē is one of New Zealand's most flamboyant birds. GLENDA REES

MURCHISON MOUNTAINS AND TAKAHĒ SPECIAL AREA

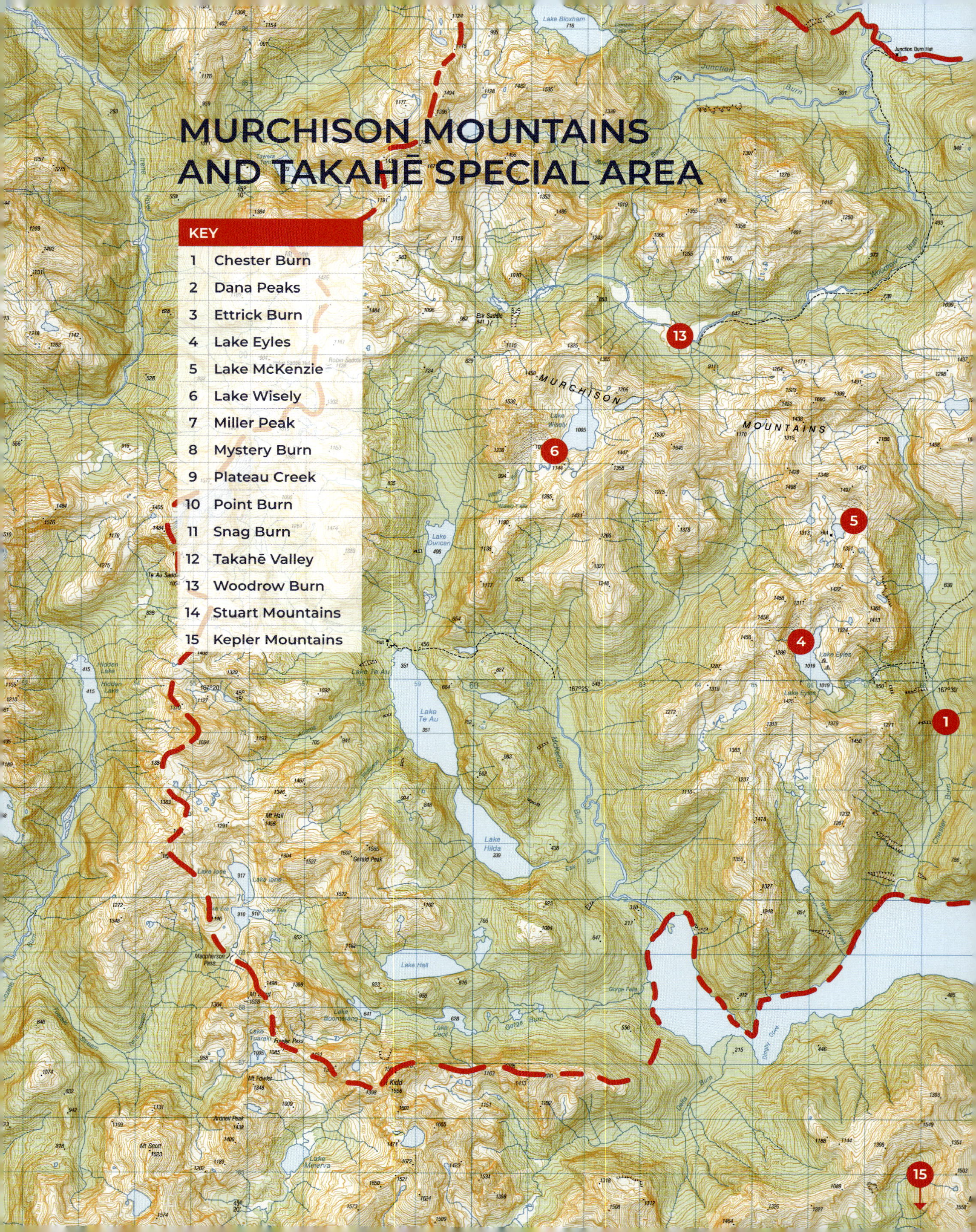

South West Arm
Middle Fiord
Arran Island
Cumbrae Island
Bute Island
Middle Fiord
Rocky Point
Manea
Erin Island
Doubtful Island
Lake
Aurora Point
Miller Peak
Dana Peaks
Hut
Log cabin
Ettrick Burn
Ettrick Burn
MURCHISON MOUNTAINS
Mt Lyall
Takahe Valley
Lake Orbell
Tunnel Burn
Te Ana-au Glow-Worm Caves
Black Cone
Mt Owen
Deadmans
Garden Point
Mark Point
Moonlight Beach
Honeymoon Beach
Moat Point
South Fiord
Fiord
Lake Te Anau
Lake Te Anau
Dome Islands
14
7
11
2
9
3
12
10
8

CHAPTER ONE

HĀPARA – A WILD LIFE

There's a faint rosy glow on the eastern horizon and the sky above the high ridges on this November morning is beginning to lighten imperceptibly, but it's still dark on the side of the valley. Under the drooping thatch of a large narrow-leaved snow tussock, a big bird shifts slightly as she uses her large adze-like bill to gently roll the two creamy eggs she has tucked between her feet underneath her warm chest. She rocks slightly from side to side as she settles herself again, tucking her head under her wing. A few metres away her mate lifts his head and listens to her move. He makes a reassuring 'oompf', a gentle thump that resonates from deep in his belly, then he, too, returns to sleep.

It's not long before the stars dim and then disappear altogether as the sky brightens. The wheaten snow tussocks shimmer and rustle in the light breeze, and both birds stir again. The incubating female lifts her head and carefully rises, leaving the eggs nestled in the soft bowl of dried grass. She steps forward through a tunnel of snow tussock and carefully pushes through a concealing screen of stalks. There is a shock of colour as her vivid red bill and frontal shield appear through the pale tussock, followed by a bulky body clad in a shining coat of deep purplish blue, shading up through dark turquoise to an intense olive-bronze on her back. She stretches her strong, stocky pinky-red legs and then moves quickly towards a latrine pile of droppings a couple of metres away. Each dropping is a miniature hay bale, a 10-centimetre-long golden-brown cigar of fibrous undigested snow tussock, and this latrine pile has nearly a month's worth of waste piled up.

Her mate appears, oompfing quietly. The pair face each other, almost touching, and lift their heads. He bends forward to briefly preen the side of her neck, before he turns, ducking slightly as he moves into the nest to take his turn incubating the eggs.

As the female moves away in search of food, she strolls slowly yet purposefully, placing each foot very deliberately. As she brings a foot up high and moves it forward, her three long front toes curl inwards towards the shorter rear toe. As she places it firmly on the

Takahē usually lay two eggs, although clutches can range in size from one to three eggs.
ROD MORRIS

ground, the toes widen outwards like a broad snowshoe. Her head is high and alert, and jerks forward with each step. She flicks her tail constantly, revealing a bright flash of white feathers. She weaves her way among the bright green vegetation of head-high hebe bushes interspersed with snow tussocks until she reaches a slight rise, where a fan of rocky debris from the hanging valley above has built up over time. The snow tussocks change species from narrow-leaved to mid-ribbed, and she begins to sample from different plants. Pushing her bill into a tussock, she takes hold of a single tiller and, rocking back on her haunches, tugs it out. Each tiller is a short stem supporting four leaves, wrapped at the base with a tough sheath, and her tugging breaks the leaves off where they join the stem. Lifting one foot, she grabs hold of the tiller and uses her bill to strip off the outer mature leaves, before delicately snipping and swallowing the bottom 2–3 centimetres of the young inner leaves. She discards the rest and moves on to another snow tussock. This one is much more to her liking, and she eats nearly a dozen tillers, discarding most of them in a pile beside her.

The female feeds like this for several hours without stopping, leaving behind her a trail of discarded tussock tillers and her distinctive hay-bale droppings, which she produces every 10 minutes or so. Over the course of a day she might leave a trail of poo that would measure 7–9 metres if it was all lined up.

Then, almost trotting, she rushes back to the nest. She enters via a back tunnel, and after a few minutes her mate appears outside; it's his turn to feed. Before she settles, the female rolls the two eggs. She listens intently – tiny cheeps can be heard from the oldest egg, as the 29-day-old chick inside prepares to hatch. The female clucks in reply and, wriggling, settles down. The pair continue to take turns incubating, and when it's nearby the non-incubating parent stays in touch with quiet hoots. By the end of the day the pipping chick has just broken through the eggshell with its sharp egg tooth. It spends the night slowly rotating, doggedly chipping a lid at the broad end of the egg. Its mother stays attentive, clucking to the cheeping chick.

As the sky lightens, the chick finally breaks free. Hāpara, whose name means 'dawn', lies exhausted between her mother's feet, a tiny defenceless 100-gram bundle covered in wet down. The father sticks his head into the nest, keen to take over incubation, but the mother is reluctant to leave. Finally, with some encouragement, she departs and the male settles over the exhausted chick. His body warmth soon dries the black down, which fluffs up in a dark halo around the striking white of Hāpara's tiny bill. For the first few hours, the chick sleeps, drawing on the last of her yolk reserves for nutrition. But it's not long before her parents are feeding her regurgitated tussock juices, a job made easier by the chick's white bill – it stands out in the dimness of the nest and enables the parent bird to accurately position its bill against hers.

By the time the second chick hatches, two-day-old Hāpara is peeping noisily when she is hungry, moving around the nest when she can, actively begging for food and starting to gain a little weight. Her brother is healthy but he struggles to compete with his larger sister for their parents' attention. The parents, which are themselves mostly vegetarian, catch small insects rich in protein to feed to the nestlings. They dig out plants and fossick in the soil underneath in search of invertebrates, ranging from spiders and worms to the larval, pupal and adult stages of many insects. They might turn over several square metres of ground each day as they search.

The family stays in the nest for the first fortnight, but by early December the chicks are strong enough to accompany their parents as they feed. The four birds stay close together, and there is a constant chatter of contact calls from the parents and high-pitched peeping from the chicks. The parents continue to dig and take turns delicately feeding the chicks tiny insect morsels and the juiciest bits of leaf stem. At the slightest hint of danger, such as a bird flying overhead, the parents cry out a loud, deep oompfing alarm and the chicks scuttle for cover, hiding until their parents sound the all-clear. At night, the family returns to the sheltered familiarity of the nest.

The white tips of a young takahē's beak make it easier for its parents to feed it, especially in the dimness of a nest. **ROD MORRIS**

The nest is in the centre of the pair's breeding territory, and their feeding forays take them from the shore of the small lake through a varied patchwork of low shrubs, flowering herbs and snow tussock meadows. Both parents and chicks happily get their feet and legs wet at the edge of the lake and splash through shallow swampy areas. But while water underfoot is one thing, water falling from above is another matter. Chick down is not waterproof, and when it rains the parents lead the chicks under large tussocks or bushes with tightly knitted canopies in an effort to keep them dry.

Fiordland weather is seldom fine for long: rain is frequent and often torrential, and can be accompanied by fierce winds. On one such day, when the chicks are little more than a month old, a severe southerly storm sweeps the valley. A 24-hour downpour is followed by a dusting of unseasonal snow and freezing temperatures. Despite the birds' best efforts to stay dry, the younger chick becomes damp and chilled, and his health soon deteriorates. Sick and unable to feed, his death comes quickly.

Hāpara is now the sole focus of her parents' attention. She is growing her first dark feathers, and as the black down is pushed out it begins to slough off against rough bushes. Her legs change from a horn colour to a darker grey. By the time she is six weeks old, her parents begin to switch from feeding her insects to giving her only the juicy bases of tussock tillers.

At two months of age, Hāpara is tall and gangly and weighs over a kilogram. The family have long since left the nest she hatched in, and instead stop at night wherever they are to seek shelter under bushes and tussocks, sometimes building temporary brood nests. When the weather is fine, their daytime wanderings take them up to the highest parts of their breeding territory, where the vegetation is low and stunted. One day, as they explore the edge of their 30-hectare territory, they notice another pair of takahē moving through low scrub on the far side of a tumbling stream. All four adults strike up a loud chorus of territorial calls, monosyllabic 'kleup' notes that each rise indignantly in pitch at the end. Hāpara peeps loudly in support but she is not yet able to make the same calls as adults. After a few minutes, the calls quieten and both sets of birds, satisfied that their territory hasn't been encroached on, return to feeding, slowly drifting away as they do so.

Another day, the family encounter a single bird, making its way down a steep gully from the ridgeline above. The visitor is interested in the family, but when the father rushes forwards assertively, kleuping loudly, it flees downhill, its head thrust forward and wings held out as it runs as fast as it can.

By the end of summer, Hāpara is four months old. At 50 centimetres tall and weighing 2.3 kilograms she is the same size as her mother. She increasingly forages for herself, although her parents continue to feed her the tastiest morsels. She learns by copying the older birds, pulling and snipping tussock tillers using her bill, which is now completely dark. It takes 15.5 kilograms of force to pull out a single tiller, so she has had to wait until

Waterfall Basin contains typical subalpine and alpine takahē habitat that is a mosaic of tussocks, flowers and small bushes such as hebes. Murchison Mountains. **ANDREW DIGBY**

she weighs as much as an adult to have the strength to do it. Her parents introduce her to mountain daisies and show her how to pull out a leaf and eat the bottom few centimetres. She learns that different snow tussocks offer differ nutrients and begins to see how these change over time.

Some of the small *Poa* grasses are fruiting, and the takahē eagerly add these seeds to their diet. They strip the seeds by running their slightly open bill up the grass stem. As autumn progresses and the birds seek out starch-rich plants to boost their fat supplies ahead of winter, they wander further afield, often overlapping with neighbouring takahē. The birds are aware of one another and move away, but there is no conflict or fighting as there sometimes is during the breeding season.

In May, temperatures drop and the first winter snow falls on the high peaks and ridges, although it doesn't settle. Hāpara's parents find small patches of the summer-green thousand-leaved fern (*Hypolepis millefolium*), whose fronds have already been frosted in the cold nights. They demonstrate how she can use her beak to dig for the fern's starchy rhizomes,

Takahē continue to feed in the subalpine zone in winter, even when it is covered in snow.
ROSS CURTIS

which have sweetened with the cold, and then carefully peel the tough outer skin to reveal the white flesh within. As the days close in there are fewer hours in which to feed, and sometimes it is after dark before the family seek shelter for the night.

By June, the high alpine zone is blanketed with snow, and on the shady side of the valley heavy frosts have frozen the ground. The family have moved to the lowest parts of their home range, where there is still plenty of food and shelter in among the mosaic of bushes and snow tussocks. An intense midwinter low crosses Fiordland, and the family emerge one morning from their night-time shelter to find their whole world is white, although there are plenty of snow-free areas under bushes and thick tussocks where they can carry on feeding. They stolidly plug across snowy ground, their wide feet mostly riding across the soft snow. When it gives way and they plunge through up to their belly, they unperturbedly extricate themselves and soldier on in search of tillers and rhizomes.

A few weeks later, there is another heavy snowfall. The temperature stays well below freezing for the following few days, and ice begins to encroach from the edges of the lake towards the middle. The wet snow turns to solid ice. The trio have already beaten a retreat into the shelter of the forest, heading to familiar areas where the parents know they'll find *Hypolepis* fern rhizomes. They patiently grub out the subterranean stems, which make up the bulk of their winter food, but they also search out rushes, eating just the base as they do with tussock. They make do with that they can find; when they can, they also eat the bases and rhizomes of several species from among the sedges, including several kinds of hook sedge and rautahi or cutty grass. Their droppings change from dry hay bales to smaller dark, moist splodges.

Through late winter and early spring the takahē drift in and out of the shelter of the forest. The lake is covered in thick ice, and the birds take shortcuts across it as they look for areas where they can still access snow tussocks. No sun has reached the valley floor for nearly three months, and on frosty nights the temperature plummets to -17°C, but it's warmer under the insulating snow and the family spend the nights under thick tussocks, huddling close for warmth.

Spring comes slowly and the snow begins to melt, further wetting the already swampy valley floor. The ice on the lake cracks loudly as air temperatures gradually rise, and the water that has continued to flow under the ice begins to move faster. Soon, the snow has melted off the bushes, although drifts continue to lie on the cold ground. The family push further and further up the valley, and although they still move together, the parents are taking more notice of each other as the days get longer. They begin prospecting under tussocks, and gather tillers to line nests that they don't quite complete before abandoning them and moving on to try another in a different location. As with everything, Hāpara copies her parents, gathering tillers and helping press them into the nest bowl. By late October, the parents are courting in earnest and there are times when this culminates in mating. They have chosen their nest site, and at intervals of two days the female lays three eggs, each the length of a human palm. The parents again take turns incubating, and this time Hāpara takes her turn as well.

A month later, two of the eggs have hatched, but although the trio keep incubating the remaining egg in the nest, it is infertile. After several weeks it is eventually abandoned when

A young takahē amongst flowering Mount Cook buttercups. **JAMES REARDON**

the family move away with the two new chicks. The new arrivals thrive with the care of both parents and extra attention from eager helper Hāpara, who is learning her parenting skills well. But tragedy strikes the little family in early January. An opportunistic weka has been attracted by the incessant peepings of the chicks, and even though the adults fight back and do their best to defend both chicks, the attacking bird is fast and determined, managing to grab one of the chicks. After that experience, the parents are extra vigilant about protecting their one remaining chick, but although they often hear the weka calling from the edge of the forest, it never comes close again. When the weka calls, the parents even call back, loudly announcing their intention to defend their territory. These weka–takahē duets echo around the valley on still evenings.

As the new chick grows, Hāpara begins to spend less time feeding it and more time on her own, feeding away from her parents. She still returns at night to sleep with them, but by the end of March she is feeling restless. One day, she starts walking. Feeding as she goes, she begins a journey that will take her several weeks. Slowly, she heads up the valley. She passes through several other takahē territories, but the pairs are feeding elsewhere and she passes by unnoticed. At the head of the valley she climbs steadily. The vegetation gets shorter and the ground rockier. There is not much to eat up here, and she forges on, enticed by the warm gusts of wind that drift across the pass. She pauses at the top of the climb, unsure which way to go. The ridge stretches away on either side, rising to rocky peaks. In front of her is a north-facing basin, and she can sense the tall tussocks growing further down the slope. She begins to pick her way down, pushing between the bushes and thick snow tussocks, pausing when she senses a tussock that is worth trying. Many of the alpine flowers she passes have been significantly chewed by deer, and some of her favourite species of tussocks have been browsed almost to ground level. Deer droppings are numerous.

She stays in the basin for several days, gradually exploring more and more of it. There is a small stream zigzagging down the middle and the haunting cries of kea flying overhead echo off high cliffs on one side. Eventually, her wanderings bring her close to the resident pair of takahē, and although there is no animosity this is not what she is looking for. She moves on, nearly running into a herd of deer heading in the opposite direction.

Hāpara's journey takes her across another high saddle to an alpine head basin perched high above a forested valley. There is a large debris fan extending below the basin, and on the slopes the snow tussocks are lush and tall. As she wanders around she notices old feeding sign scattered on the ground, and then, on the far side of the small stream, she begins to come across fresher signs of browse: piles of tillers and fresh green droppings. It's not long before she encounters the cause of all this sign. The bird is on his own and he regards the stranger with interest. They are curious but cautious, and they keep an eye on each other as they wander and feed.

The other bird, Porehu (meaning 'dusk'), is a widower. He is 10 years old and the father of two grown chicks. His mate drowned in a flooded stream a few weeks earlier, and as their only chick had died early on in the breeding season, he is unencumbered and on the lookout for a new mate. There are other takahē in the valley, but they are already paired up.

Slowly, Hāpara and Porehu move nearer, eventually getting close enough to introduce themselves. They face each other and lift their heads. Porehu steps forward until their chests touch lightly. He calls quietly and Hāpara replies. A bond is cemented.

The pair are soon inseparable. Porehu knows this valley well, including where to find the juiciest tussock tillers, the most succulent mountain daisies and the crunchiest *Hypolepis* rhizomes. Side by side, with a constant chatter of contented oompfs, the pair feed well in readiness for the winter. If they wander too far away from each other as they feed, they make a loud kleuping call, to which the other duets a reply.

One day, as they feed out in the open on an old slip, they notice a strange animal standing a few hundred metres above them. It is tall, and as it slowly comes down the hill towards them they see it has two legs. They are as curious about the stranger as it is about them, and they watch its approach. It is a human male, the first they've seen. He wears a long woollen Swanndri bush shirt, and he has a pack on his back and a rifle slung over his shoulder. 'Hello tarks,' he mutters, as he reaches into a pocket for a small notebook and pencil stub to make a note of his sighting. As he stands there, the birds drift closer. There are strange smells: wood smoke and the acrid odour of gunpowder. Hāpara and Porehu wander off the slip and into a maze of bushes, and they hear the rattle of rocks from behind as the man turns and climbs up to the ridge.

Half an hour later, there is a loud gunshot as the man shoots a deer, but although the birds startle with the sudden noise and its echoing, they don't link the strange animal with the sound. For them, it is in the same category as thunder or a rockfall – just another part of the local soundscape.

Hāpara passes the winter with Porehu, discovering the *Hypolepis* groves in the forest and making her way in the myriad tunnels under the snow and bushes on the low sides of the valley. In late spring, their courtship starts in earnest. They touch and preen each other and duet seducingly. They run together, then stand close, facing each other with their necks stretched up, circling and gently pecking their partner's neck. When they are not courting, they search for a nest site and build a few trial nests. When she is ready to mate, Hāpara solicits Porehu, turning her back to him, spreading her wings low and allowing him to mount her. She lowers her head to the ground to stabilise herself as she holds his weight. He keeps his feet spread and his legs bent to make himself as low as possible, and spreads his wings to help him keep his balance as they touch cloacas for a few seconds. If he doesn't accidentally overbalance and fall off, he finishes a successful copulation by stepping forward over her head. Each mating is brief, but over the course of a few days they mate many times.

Although it is her first clutch of eggs, Hāpara knows the incubation routine from her time the previous summer as a helper, and Porehu has many years of experience already. She mostly incubates during the day, and he takes the overnight shift. But 30 days come and go, and still no chicks hatch. The pair begin to get restless, although it's two weeks before they abandon the infertile eggs. After a break, they begin courting again and building trial nests, but nothing comes of it.

A year after his last visit, the man is in the valley again. Although the birds see him, they are well hidden among the tall tussocks and he doesn't see them. Other people appear in the lower valley one day, and there is the sound of axes and hammering as they clear a small area of scrub. Soon, there is a loud noise from the sky and the takahē look up to see a large flying object, bigger than any bird they have ever seen. They hide under a bush as the helicopter roars overhead, a small building suspended below it. The helicopter hovers for a few minutes above the new clearing, lowering the portable bivvy onto the ground. In the silence after the helicopter departs, there is more banging and noise as the men establish wire guy ropes to attach the bivvy in place.

A few weeks later, Hāpara and Porehu's travels take them near this new structure. It smells of freshly cut wood and metal, and they walk around it for a while, getting accustomed to it. Soon, it becomes just another part of the valley, mostly empty but sometimes occupied

Botanist Bill Lee in the 1970s, marking out a vegetation transect through tussock in the Murchison Mountains. **WILLIAM LEE**

by one or two men, who notice the fading droppings that the passing birds have left on the ground around the hut.

The next summer, when Hāpara is three years old, she lays two eggs, one of which hatches successfully. She and Porehu are diligent parents. For the first few weeks of the chick's life they excavate grubs and larvae, feeding it constantly. It trails alongside them, peeping, as they cluck reassuringly in return. When they notice three men in the valley, they shepherd their chick cautiously away, but over the next few days they relax as they grow accustomed to the presence of people. Tall metal stakes appear in the ground at regular intervals, marking out areas of different tussock species, and tape measures lie stretched across the plants from stake to stake. The men spend long periods in these areas, counting the piles of tillers the family have left behind as they feed, and cutting and pulling tillers of their own. The takahē family are faintly curious and pause their activities at times to watch and listen to the men talking. In their turn, the men observe the family, which is out feeding each day from first light until the long solstice dusk finally darkens around 11 o'clock each night. After just five hours' rest, they begin another 19-hour feeding marathon. It is a hot summer, and although the valley lies 1,000–2,000 metres above sea level, the usually cool air temperatures climb above 20°C for days at a time.

The men disappear for a few weeks, but in late summer they return. By now the chick, Tahi, meaning 'one', is a leggy, inquisitive adolescent. He confidently feeds a few metres away from his parents, but he responds quickly if his parents oompf a loud warning, dashing under the nearest bush. The parents sense danger from the shadow of a kea flying overhead and the repetitive 'yak yak yak' of a New Zealand falcon hunting on the forest edge. The humans, however, don't signify danger, and at times the family comfortably feed nearby as the men work. One day, Hāpara drifts close to one man who is kneeling and concentrating on a tussock bush. The man hears a deep, quiet thump, and when he slowly turns it is to discover Hāpara's face, close to his own, peering over his shoulder. He smiles, stay still and watches as she wanders away back to her mate. 'You're a friendly pair,' he says quietly, as he returns to his task.

Early one morning the mountain peace is interrupted by the loud 'whack-whack' of a helicopter, which bursts into view over a ridge. A man peers out through an open door, lifts his rifle to his shoulder and, as the machine hovers above the head of the valley, fires off a few rounds of ammunition. Four deer drop to the ground. A minute later, the helicopter has landed nearby and the shooter leaps out to collect the carcasses. Job complete, the helicopter lifts up into the air and swiftly drops down towards the valley. As the blast of noise passes overhead, the family duck under a nearby bush, then as the sound quickly recedes into the distance, they venture out to carry on feeding.

Come autumn, the birds discover an unexpected bonus. In the area where the men were working, they cut some tussocks back heavily as part of an experiment to simulate the effect of deer browsing. To the delight of the takahē, the tussocks have had a growth spurt and the newly sprouted leaves are tender and sweet. They gorge on the regrowth.

Winter arrives. Temperatures start to drop and the days begin to shorten. Snow blankets the ground and the family has just 10 hours of daylight to feed in. When storms blast the mountains, the birds seek shelter and *Hypolepis* in the forest. Then when conditions permit, they return to the lowest patches of scrub to search for *Hypolepis* and accessible tussock there.

In spring, Tahi stays with his parents as their helper, the three of them taking turns to incubate two eggs. Two chicks hatch – Rua and Toru, meaning 'two' and 'three', both females.

This year, things are different. The previous summer's record temperatures have triggered a mega-mast, a mass flowering and seeding of all the *Chionochloa* tussock species in the alpine zone, as well as the beech in the forests. In spring, the forest canopy is red from the myriad tiny flowers. When the wind blows, waves of pollen are swept from the beech flowers like great yellow dust clouds. The multitude of flowers provides a floral banquet for many forest invertebrates, whose numbers begin to climb. By early summer, a carpet of spent flowers covers the forest floor. In the alpine zone, flowering spikes are developing among the tussocks' leafy tillers.

The mid-ribbed tussock (*Chionochloa pallens* subsp. *cadens*) often grows on debris slopes in the Murchison Mountains and is a favoured takahē food. DOC

The men return, on occasion, to the valley, to count and measure. They notice from the discarded stems that the takahē have been focusing on the flowering stems as they pluck tillers and nip off the juicy bases. As summer progresses, the birds turn their attention to the prolific seedheads that begin to wave above the grassland like massed flags. Hāpara and Porehu show the younger birds how to gently grasp a flower spike and run their beak towards the top, gathering up the generous bounty into their gape. For weeks they feast on the seeds. And in the early autumnal forest, the first of what will be millions of tiny beech seeds begin to fall.

The takahē are not the only birds enjoying the seed extravaganza. In the forest, kākāriki and kākā are also having bumper breeding years. The abundant seed arrives in time for later clutches of kākāriki chicks to all survive. Large family groups feed noisily in the canopy, the passage of the parakeets marked by their chuckling chatter and the rain-like patter of discarded seeds falling to the ground. Many pairs are in such good condition that they will breed again, the ongoing harvest of seeds feeding them well into winter.

The kākā, meanwhile, are relying on the bonanza that they've known was coming for many months. They don't breed every year, but this year all the local pairs have laid eggs by the end of January, and there is plenty of food to see them through the next few months of chick-rearing. The exuberant, raucous cackles of these large forest parrots ring out as they forage up and down the valley.

But a sinister chain reaction has also been set in motion by the abundance. Mice are feeding on the masses of seeds on the forest floor, and are starting to breed. They have large litters of 10 or more babies, which rapidly reach maturity and begin to breed themselves. By early winter their numbers are rising rapidly, and they will keep increasing through the winter, making the most of the seeds that still litter the ground. The growing mouse population, in turn, becomes a new food source for the ship rats, which are beginning their own population boom.

This rise in rodent numbers is a boon for the local stoats, which find plenty of easy prey. The female stoats are all already pregnant, and the rich supply of rats and mice through the coming winter will ensure the developing embryos all survive.

In the meantime, there are changes afoot in the takahē family. Tahi has become restless, and soon it is time for him to leave in search of a mate and a territory of his own. It turns out that he doesn't have to go far. When he encounters the neighbouring pair, the female takes an instant shine to him. She has been with her mate for a few years and they have failed to raise any chicks. Her partner is not impressed with his mate's shifting allegiance and he starts a fight with Tahi. But he is an old bird and Tahi is a big, strong youngster, and the fight does not end well for him. Sporting several minor injuries from the tussle, he beats a retreat, leaving Tahi to his new beau.

Hāpara and Porehu are oblivious to the game of musical partners going on next door, as they are busy with Rua and Toru. They continue to enjoy the plentiful snow tussock seeds, until the winter snow starts to arrive and eventually buries the remaining seedheads.

This year, when the takahē family retreat to the winter forest, they find a growing plague of rats and mice, but it has no impact on them … for now.

In spring, just as their populations reach a peak, the rats and mice begin to run out of food, as the beech seeds have either germinated or rotted on the ground. Rat numbers are at a record high – more than 17 rats per hectare of forest – and the hungry rodents are growing desperate for food, even as a new cohort of young stoats begin to feast on them.

This breeding season five-year-old Hāpara has two teenage helpers as well her loyal mate. In late October, she lays three fertile eggs, and with the hard work of incubating spread between four birds they have a relatively easy time of it. But they are not the only family in the neighbourhood. A female stoat has nested under some thick bushes nearby, and in early December she gives birth to 10 kits. With 10 hungry mouths demanding to be suckled, she has an appetite that is even more voracious than usual. She is indiscriminate about what she eats. Mice feature largely, as do ground wētā. Rats and small passerine birds are popular, too.

The birth of the stoat kits coincides with the hatching of the first takahē chick, followed two days later by its sibling. The third chick has just hatched and is still wet when the

Introduced stoats thrive in both the forest and in the subalpine zone. **NGA MANU IMAGES**

stoat attacks under the cover of darkness. Emboldened by the smell of the newborn and undeterred by the presence of Porehu, who is 10 times heavier than her, the female stoat rushes into the nest. In the ensuing struggle, the newest chick is crushed before Porehu can fight off the predator, but miraculously the other two survive. Vanquished for now, the stoat retreats. But not for long.

Driven by her insatiable desire to kill, the stoat returns the next day. This time, Tọru is incubating, and although Hāpara and Porehu have remained nearby, the stoat makes its first move before they have a chance to react. This time the stoat enters via another tunnel and comes upon Toru from behind. Acting reflexively, the stoat delivers a killing bite to the nape of the bird's neck. The frantic peeping of the chicks alerts their parents, and they rush to the nest, only to find themselves face to face with the stoat, which now has one of the chicks dangling from her mouth. This time Porehu has an advantage – he is standing and he towers over the stoat, which tenses, ready to leap. But she is a split second too slow. Like a frenzied kick boxer, Porehu lashes out with a foot, kicking the stoat backwards. With his other foot he rakes at her neck and his long, strong claws pierce her spinal cord. He stabs at her with his adze-like bill, nearly decapitating her. In less than a day, the takahē family has shrunk from seven to just four. Hāpara, Porehu and Rua gather up the one remaining chick and usher it away from the carnage. They will build another nest to shelter the little one for the next few days until he is strong enough to join them on their wanderings.

In the stoat nest, it doesn't take long for the 10 kits to starve to death when their mother fails to return. But many more stoat nests succeed that summer, and by January the stoat population in the valley is boosted from its usual half a dozen or so to more than 50 animals. It's a predatory onslaught and no creature is safe. As rat and mice numbers crash from lack of food, the stoats, too, become increasingly desperate. All the small birds, from rock wrens in the alpine zone to brown creepers in the forest and whio in the river, are targeted in the stoat killing frenzy. Few takahē chicks will survive this summer, and Toru is not the only adult to die.

When the botanists return to count their vegetation plots, they notice the resident 'friendly pair' and see it has a chick as well as a helper. They feel reassured, although they comment among themselves about seeing fewer birds overall and worry about the reason for this.

The deer-cullers certainly aren't finding as many deer, but that's a good thing. The helicopter returns and searches the valley intently, but the shooter fires only twice.

There is no windfall of tussock seed this year, although there is still plenty of food on which to raise Whā, chick number four. As summer draws to a close, it's Rua's turn to leave home and find a mate. With so many adult takahē deaths this summer, it won't take her long to find a male on his own.

Winter descends and with it comes a slow end to the stoat carnage. The boom months are over for the predators; a few months after the rat population crashes, stoat numbers also plummet. By spring, equanimity in the valley is restored and it's time for the remaining birds, takahē and otherwise, to get on with life and slowly rebuild their numbers.

Back in the valley, the men arrive early in spring and this year they are doing something different. They have marked out small areas of tussock, and after some intense counting of tillers they begin walking through each plot spreading different kinds of fertiliser. An acrid smell drifts on the light breeze.

Hāpara, Porehu and Whā have survived the winter but this will not be a good breeding season for them. The two eggs are infertile, and with no chick-rearing duties to occupy him, Whā leaves home earlier than his siblings. His journey is a long one and he crosses several catchments before deciding to settle in what appears to be an empty territory. It will be another year before he finds a mate.

It is a cool, damp summer, and with no chicks to care for Hāpara and Porehu range widely across their territory. On a few occasions they run into Tahi and his mate, feeding near the edge of their territory. They are also chick-free, and despite the family connection the pairs react assertively, calling loudly to ensure that the other birds recognise where territorial boundaries lie.

The underground rhizomes of the summer-green fern *Hypolepis* are an important winter food for takahē. Chester Burn, Murchison Mountains. PETER MOORE

In late summer Hāpara and Porehu are back in the area of the fertiliser plots. In one plot, their favourite mid-ribbed snow tussock is much lusher and darker green than usual, and they linger here for a while, feeding intently.

Winter comes and goes, and the pair endure the cold weather as usual, moving into the forest when the weather is very bad and heading back into the valley when things improve. The spring thaw finally comes, and as the snow recedes the birds move up-valley again.

Porehu is now 16 years old, which is a ripe old age for a Fiordland takahē. Although he courts and mates as enthusiastically as ever, once again he fails to father any chicks. He survives the summer, but his health is beginning to deteriorate, and in early winter he dies, leaving seven-year-old Hāpara on her own. She survives the winter alone, but takahē prefer the company of others, and in late spring, when a lone male appears in her territory, she loses no time in introducing herself. He is available. She is available. That's all takahē need to have in common to strike up a bond. Kōanga, meaning 'spring,' is a five-year-old whose mate was killed a few months earlier in a snow avalanche. Although they had bred, none of their chicks survived past a few weeks old.

Kōanga is a banded bird. He has a metal bird band on one leg and two coloured plastic bands on the other. These were put on by researchers several year earlier as part of their regular survey project to find out how many takahē there are and how well they are breeding.

Hāpara and Kōanga waste no time, and by mid-January they are the proud parents of two chicks. However, the youngest chick struggles to compete against its older sibling and dies at just two weeks old. The female, Rima, thrives under Hāpara's experienced care, and Kōanga, who hasn't raised older chicks, soon gets the hang of it. But the winter starts early, and the weather is brutally cold. Heavy snow falls and then freezes to thick ice that doesn't melt for weeks at a time. Even the forest floor is frozen and the family struggle to find enough *Hypolepis* rhizomes to eat. The combination of lack of food and constant sub-zero temperatures eventually proves a lethal one for Rima. While it is also difficult for the parent birds, they struggle on and manage to find enough food to make it to spring.

The pair mate and nest as usual, but the harsh winter has taken its toll on them and the two eggs they lay are both infertile. But they are unaware of this and diligently continue incubating. They are three weeks into incubation when a man and a woman appear in the valley below. It's Hāpara's turn to feed and she watches the newcomers as they slowly cross the floor of the valley and then begin to climb up the sides. They are walking a few metres apart and carefully watching the ground, looking for takahē feeding sign, of which there is no shortage. They head in the direction of the nest. 'I've found a latrine pile,' says the woman.

'Well, have a good look under the closest tussocks,' replies the man. 'There'll be a nest nearby.'

Wildlife Service ranger Nick Torr weighing a takahē egg in the Murchison Mountains in the mid-1980s. **NICK TORR**

The woman bends and begins to sweep the trailing tussocks to one side. The next moment she and Kōanga are face to face.

'I've found the nest,' reports the woman.

Even as she is speaking, Kōanga beats a hasty retreat out the back of the nest. Being handled when he was banded has made him wary of people, unlike Hāpara, who retains her naive fearlessness.

The people crouch by the nest, and as the woman puts her head inside a large black cloth bag the man lifts one egg out to pass to her. She slips the hand with the egg through a short sleeve and inside the bag. She turns on the torch in her other hand and holds its beam against the egg to 'candle' its development. Inside the dark bag the egg glows as the torch beam shines through it.

'Oh. It's infertile.'

They candle the next egg and the news is the same: infertile.

The two rangers move away from the nest and watch for a while, trying to see the birds' legs so they can identify them. But Hāpara and Kōanga keep their distance and their legs remain hidden among the foliage. When the rangers finally depart, they take the infertile eggs with them, leaving an empty nest for the takahē when they return. The birds are a little nonplussed, but it isn't long before they abandon the nest completely. They are in no fit condition to try breeding again and return instead to their usual routine of walking and feeding.

The following winter is milder and it is less of a struggle to find food, so the pair are in better condition by the time spring arrives. Ten-year-old Hāpara lays two fertile eggs this time, and the incubation is going well. The two rangers reappear when the eggs have been developing for just over three weeks, and they go through the same process of finding the nest and candling the eggs. This time, though, they return both eggs, and after they have left the nest the birds soon return to carry on incubating. The rangers are watching through binoculars from a distance, and they manage to make out Kōanga's band combination as well as seeing that Hāpara is unbanded. When they check their records later they will realise that Kōanga has moved quite some distance from where he was banded and they will speculate about the fate of his first mate.

A few days later, Hāpara and Kōanga hear a helicopter making its way up the valley. It lands a couple of hundred metres away, and a man hops out and heads directly towards them. Kōanga is feeding nearby and he melts away into the bushes to hide. As the man nears the nest, Hāpara quietly slips away as well. The ranger soon finds the nest and reaches for one of the eggs. He nods, reassured, when he feels that the egg is warm and he tucks it carefully into a padded container. He returns to the waiting helicopter with the precious fertile egg and climbs aboard. The two takahē stay hiding as the machine's rotors noisily wind to full speed, creating a strong downdraught that flails the tussocks, until suddenly it rises, turns and speeds across the valley, finally disappearing from sight beyond the far ridge. The silence is broken by a distant takahē call as another bird reacts to the disturbance overhead. A few minutes pass, then Hāpara makes her way back to her single remaining egg. She tucks it under her brood patch and settles down to incubate. A couple of days later, a male chick hatches, and Kōanga and Hāpara lavish it with their usual care. Ono is Hāpara's sixth chick and he thrives, moving on from invertebrates to tussock tillers and then, as winter arrives, to *Hypolepis* in the forest.

The following spring, Ono takes his turn alongside his parents to help incubate two more fertile eggs. The rangers return to check the eggs, and just as in the previous year, they remove one of them for the captive breeding programme. The remaining egg hatches, and the three older birds fuss over the young female, Whitu. It is a hot summer and the family spends a lot of time in the damper areas of their territory, enjoying the shallow pools. When winter arrives, the four birds remain together, but come spring Ono departs in search of a mate. As he leaves, the southern beech trees are red with flowers that are

A young wild-born takahē explores beech forest in the Point Burn, Murchison Mountains.
GLEN GREAVES

already shedding vast clouds of pollen, a signal of the mast to come.

Hāpara is now 12 years old. She's beginning to age, and although she lays two eggs, one is infertile. The rangers leave the family with their one fertile egg, which hatches successfully. But despite the usual care from Whitu, Hāpara and Kōanga, the young female, Waru, lives for only a couple of months before succumbing to death by misadventure: an unstable rock falls on her leg, injuring it, and she dies a few days later from a bad infection.

By late summer the takahē are indulging in a feast of snow tussock seeds, while in the forest a rain of beech seeds is falling and the rats and mice are beginning to breed. Millions of beech seeds per hectare of forest will result in nearly a thousand rats in the same valley, and just as in the last mast year, the cornucopia of rodents will result in a stoat population explosion the following summer.

It's a cold spring, and Hāpara and Kōanga breed a little later than usual. Nonetheless, when the rangers make their annual egg check, they find that the pair and their trusty helper, Whitu – who has stuck around for another year – are incubating two eggs, only one of which is fertile. The summer goes well for the family. The egg hatches successfully, and the young female, Iwa, is a strong, independent chick that blossoms with all the attention. It is now Whitu's turn to strike out on her own, and she begins to make her way out of the valley, in the same direction her brother Ono took a year earlier.

Everything seems to be going well … except for the danger from stoats. They are a ticking time bomb. The bumper populations of rats are finally dying and the mustelids are getting hungry. And as their hunger increases, so does their desperation. All the small birds are suffering the depredations of the predators, and this year size turns out to be no protection for the takahē. Iwa is the first to die, in late autumn. Her parents try to attack the young killer, but he is too fast for them. They chase him away, but later that night he returns to feed on her corpse.

In midwinter, when the first thick snows have chased Hāpara and Kōanga into the forest, disaster strikes again. This time, it is Hāpara who is attacked, and although her mate rushes to her defence, he is too late. Hāpara tries to fight off her attacker, but she is not fast enough or strong enough and the stoat, despite being much smaller, manages to deliver a killing bite.

This year, tragedies come in threes. However, what kills Whitu is not a stoat, but a slip on icy rock that sends her tumbling down a rocky bluff. The following spring, the rangers will find her body and add her to the growing list of accidental deaths. They won't find the bodies of Iwa and Hāpara, but the birds' absence will be noted. These are not the only losses. Across the Murchison Mountains, takahē young and old have succumbed to stoats. But although people will wonder about the reason for the population decline that year, they don't believe that a stoat is capable of killing an adult takahē and it will be more than two decades before that realisation strikes home.

A pair of takahē pick their way through a subalpine garden. Murchison Mountains. ROSS CURTIS

In her 13 years, Hāpara produced 10 wild-raised offspring that survived more than a few days, but only four of those survived to two years of age and had the potential to breed. Tahi and Rua remained in the Murchison Mountains. Whā died in a flooded river. Ono and his mate were caught and relocated to the Burwood Takahē Centre as one of half a dozen breeding pairs; their offspring would go on to be part of the first effort to establish island satellite populations. Two of Hāpara's fertile eggs were hatched in captivity and reared at Burwood; these birds were Ono's sister and Whitu's sister, and they were released into the nearby Stuart Mountains as part of an ultimately unsuccessful programme to establish a new Fiordland population.

* * *

Hāpara and her mates and offspring were not real takahē, but 'average' birds living an average life in the Murchison Mountains from the early 1970s to the mid-1980s. Their experiences included real-life events as well as realistic, albeit fictional, ones. These years were particularly significant, and included a period of groundbreaking ecological research and the introduction of helicopter deer-culling. They were also a time when the takahē population in the Murchison Mountains plummeted to a worrying low, kicking into motion a new era of intense field management.

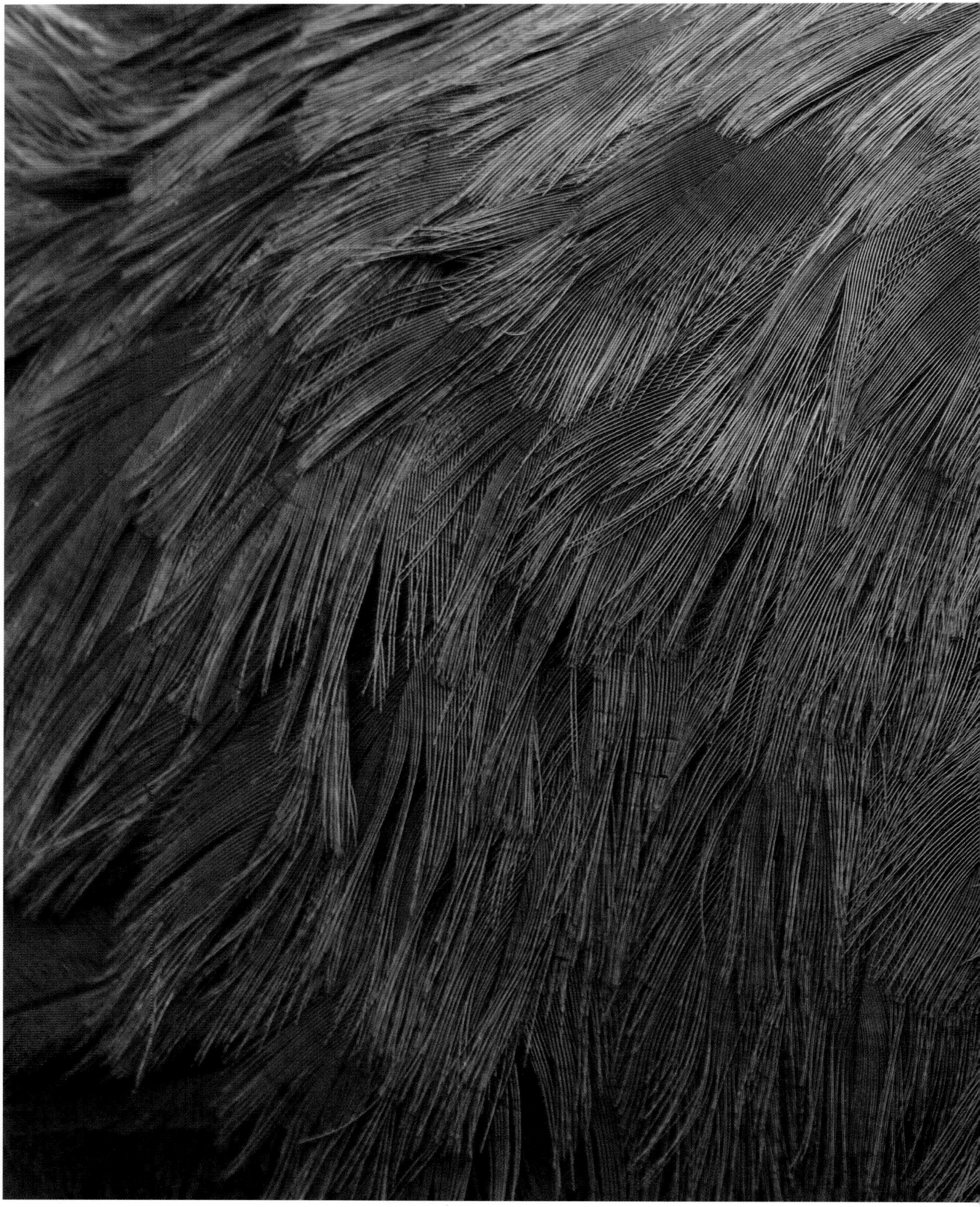

PART 1

DISCOVERY

1948 to the 1960s

Ka tū te moho, kia ora ake anō.

The takahē stands, in order to live again.

CHAPTER TWO

BIRD OF DREAMS

I'm in a helicopter with two takahē rangers and a dog, heading across Lake Te Anau. The township of Te Anau shrinks behind us as we head west. Ahead of us lie the eastern flanks of Fiordland and, to the left, Mt Luxmore and the Kepler Mountains, traversed by the popular Kepler Track walk. Hidden Lakes is a low wallow of hummocks and water at the entrance to the South Fiord of Lake Te Anau, while rising from the inlet's northern shore are the Murchison Mountains.

The steel-blue water below is calm, a mirror reflecting a low ceiling of uniform grey cloud above. And that cloud draws a straight line halfway up the mountains. The lower slopes are goosebumped with deep-khaki forest, vertically sliced in places with the earthen scars of old slips. There is no sign of the mountain tops in the cloud. We are hopeful but uncertain – if this cloud is as thick as it looks, we won't be able to land. Instead, we'll have to turn around.

As we draw closer, I see sheer rocky bluffs rising from the forest. White sandy beaches scallop the lakeshore below. I can make out the Tunnel Burn, which has carved a deep zigzag as it flows down the steep mountain flank. The stream is a portal to a hidden world, one that is well obscured by cloud.

But suddenly, just as the helicopter pilot had hoped and predicted, we see a tiny miracle: a glimpse of sunshine, blue sky and a snowy peak framed in the distant 'V' of the valley. We smile with relief. The helicopter lifts us higher, and that window of blue grows bigger.

The canopy of the forest gets closer and we can make out individual trees below. There are a few pale skeletons, but most are rich green, their branches held out to the side in the horizontal layers that are so distinctive of southern beeches. A distinctive band of creamy-yellow limestone cliffs, streaked with black, thrusts above the stream where it levels out.

Then we are above the clouds and in another world, a world of blue sky and sunshine. The valley floor widens into a rich mosaic of bright green and reddish-brown shrubs, tawny grasses and a meandering boulder-filled stream. A few seconds later, we can see the bright

Looking up Takahē Valley from the outlet of Lake Orbell, where takahē were rediscovered in 1948. **ROD MORRIS**

blue of a long lake. At the far end of the valley, well beyond the lake, is a forbidding circle of rocky ridge, a rampart well covered in late-spring snow.

The pilot lowers the helicopter gently onto the tussock flats near the lake outlet and powers off. The engine roar stops abruptly and the whirring blades slow. We have arrived in Takahē Valley.

We clamber out of the helicopter into the mountain stillness, which is broken by a distant takahē call. It's a single, short shriek, regularly repeated every second or so. A second takahē responds, its urgent replies a counterpoint to those of its partner. The call is quite high-pitched, perhaps an octave above middle C, and it's rough and harsh. After less than a minute the duet fades away – the pair have expressed their disquiet at being interrupted by the noisy helicopter and now there is feeding to be done.

* * *

On 20 November 1948, a team of four made the same journey into Takahē Valley, but they did it by boat and then on foot. They were hunters and had a quarry in mind, but instead of rifles, they carried 50 metres of fishing net, a small movie camera and a stills camera.

The team was led by Invercargill doctor and keen tramper Geoffrey Orbell, known as 'Doc'. With him were three young adventurers: Neil McCrostie, Joan Telfer and Rex Watson, Joan's husband-to-be. The four set off from Te Anau in Doc's boat at 3.30 am, arriving at the mouth of the Tunnel Burn soon after the sun rose. They tied up near the Te Ana-Au Glow Worm Caves, which had been discovered just a few months earlier. Joan remembers that it was a beautiful morning. She later wrote that 'a most spectacular scene lay before us as the vivid lights of dawn crept through. Only the indigo mountains in the distance separated the flame-coloured sky from its reflections'.[1]

Geoffrey was in search of the bird of his dreams. For decades, he had been captivated by the idea of a flightless bird that many said was almost certainly extinct, as none had been reported since 1898, half a century earlier. Here's Geoffrey, who was born in Dunedin in 1908, describing how his interest was piqued as a child:

> I was looking through some old photographs belonging to my mother – a keen amateur photographer -when I came upon a print of a bird in a glass case. My mother explained that it was a picture of a *Notornis* specimen in the Otago Museum, and that only four of the species had been found and it was now supposed to be extinct. That word 'supposed' stimulated my boyish sense of adventure. I read all I could find about *Notornis*, and as I grew older the call of the outdoors grew strong.[2]

The young Geoffrey grew up to train as a medical doctor, with a passion for boat-building, deer-stalking and pig-hunting. When his hunting took him to Fiordland, where all the previous sightings of *Notornis* – the takahē – had come from, he kept an eye out for a blue bird the size of a goose. But the only flightless birds he came across were weka and kiwi.

After graduating from the University of Otago and then working and training overseas, the newly minted eye, ear, nose and throat specialist moved to Invercargill to live and work. Fiordland was on his back doorstep and he was a frequent visitor, soon building a holiday house in Te Anau to use as a base.

Here's Doc again:

> *Notornis*? *Notornis*? The question was always there but rarely spoken of and then only jokingly or as a bait thrown to catch some story … Studying the map [of Fiordland, that hung on the wall at the holiday house] and thinking of the birds gradually crystallised a certain fact in my mind.

Te Puhi a Noa

The original Māori name for the mountain range in Te Rua o Te Moko / Fiordland where takahē made their final stand is Te Puhi a Noa, 'the topknot of Noa'. The range was later named the Murchison Mountains by surveyor James McKerrow, after celebrated Scottish geologist Sir Roderick Impey Murchison. Murchison was one of the founders of the Royal Geographical Society and a long-serving director-general of the British Geological Survey, and was first to describe the Permian geological period. Despite never visiting New Zealand, he has a glacier, a township and a mountain – as well as an entire mountain range – named after him, along with more geographic features in many other countries.

> By plotting all references to the bird on the map it became apparent that there was a vast area of Fiordland surrounded by reported instances of *Notornis* being found or seen. That piece of country was the least explored of the whole Fiordland National Park. I heard stories of a large lake seen from the air. Maori history suggested its existence and also its whereabouts.

The area Doc had in mind was the Murchison Mountains, known to Māori as Te Puhi a Noa. The mountains are in close proximity to Te Anau, being visible across the lake from the township, but in all other respects they were – and still remain – a remote and secret world. The lake was apparently known to Māori as Te Wai-o-Pani or 'the lake of the friendless' Geoffrey was always an enthusiastic deer-hunter, and was a founder member of the New Zealand Deerstalkers Association, which was formed in Southland in 1937. By all accounts he was a natural leader, and by 1948 he had become president of the association. In April of that year he decided to mount an expedition into the Murchison Mountains, with the joint aim of shooting deer and looking for takahē.

He took with him two young men, Neil McCrostie and Rex Watson. Doc led the local Sea Scout group in Invercargill, and he had known Neil and Rex since they were teenagers when, along with other Sea Scouts, they had helped him build a ketch. The young men then joined Doc on various boating expeditions in the ketch, until it was later sold, and also on hunting and tramping trips.

Doc later recalled their Easter 1948 trip:

> It was very tough going and took over four hours' climbing to get to the open tops. We climbed a ridge and suddenly found ourselves on the edge of a terrific precipice. Far below us a large lake glittered in the sun, and a valley extended three miles beyond it. This valley was covered with snow grass, which grows to a height of 4 ft. in thick, closely-set clumps. Through the glasses I located a young stag almost directly below us, but the cliffs were so high that he could only just be seen with the naked eye.
>
> We scrambled round great overhanging bluffs and rocks, and eventually managed to find a route to slide and scramble down to the valley floor, where the billy was boiled and we separated to stalk roaring stags. I shot a poor fourteen-pointer and sat down to wait for my two friends.
>
> I was almost halfway up the valley above the lake, dozing in the snow grass, when I heard a bird-call which I did not recognise. Two long deep notes, repeated twice.

The men would later say the calls sounded as if someone were making a whistling noise by blowing across an empty .303 cartridge case. Doc again:

> I was very tired and I sat and wondered about them. It was mid-afternoon and I was in no frame of mind to investigate noises while running the risk of being benighted in the bush. Rather than attempt the awful climb out of the valley we decided to push on and try to get out by way of the gorge below the lake. On a beach at the lower end of the lake we found fresh bird tracks large enough to make quite certain they were the tracks of my dreams. The sun was already sinking behind the mountains. We had to hurry to get down through 2000 ft. of bush before dark, so I measured the tracks as carefully as possible by scratching marks on the stem of my pipe, which is always in my mouth, and therefore less likely to be lost.

Were these the tracks of the mysterious *Notornis*? Doc wanted to check with experts. On his return to Invercargill he drew what he had seen, measured to scale against his ubiquitous pipe, and mailed the drawing to Brian Marples, a professor of biology at the University of Otago and president of the Ornithological Society of New Zealand. While he waited for a reply, Doc checked all the books he could find and became convinced that what he and the others had seen matched some measurements taken by nineteenth-century naturalist and collector Walter Buller. But the professor professed the tracks were most likely from a white heron, while bird expert Robert Falla at the Dominion Museum (the predecessor of the Museum of New Zealand Te Papa Tongarewa) couldn't match them to any known footprints.

When Doc read in the newspaper that another party was planning an expedition into the Murchison Mountains to find the takahē, he was galvanised into action. Joan, Rex and Neil were keen to accompany him, and so the party found itself at dawn that November day on the shore of Lake Te Anau, ready for the climb into the valley. Doc recalled:

> The trek was long and tedious, up terraced slopes, through dense beech forests, twined with lawyer; the undergrowth was a tangled mass of roots, ferns and rotting trees mainly obscured by moss – these conditions on numerous occasions checked our progress.
>
> After three-and-a-half hours' climbing in the bush, we finally emerged onto a clearing covered with snow grass and boxwood, and further on, snow grass only. As this flat was being crossed, it was noticed that the snow grass had been pulled out at intervals in the same manner as a swamp hen pulls reeds. These places also showed many droppings suggestive of a large bird. A few minutes later I saw the first *Notornis*. My first reaction was surprise at the small size of the bird. Thereafter our plan of action moved steadily to its successful conclusion.

Geoffrey knew that he and his team would need strong evidence to prove their discovery, which is why he was armed with a movie camera and a stills camera. Remembering that day, nearly 70 years later, Joan recollected that Doc told everyone to move slowly and quietly, and to communicate with hand signals only. Doc saw the bird first and they all crouched, then Joan said that she caught a glimpse of a bright red beak as she peered through the thick snow tussock. Doc prepared to shoot some film:

> By careful stalking through the tall snow-grass I was able to take a full reel of colour movie-film through the telephoto lens. Then the arrival of a second bird on the scene supplied the interest for another half-reel from a distance of 20 yards. When I stood up the birds were not more perturbed than hens, and only moved a few feet into the snow-grass, where the net was quietly circled around them. They walked into it of their own accord.

Joan looked down at the birds and thought, 'I have to do my thing!' She reached down and grabbed one of the takahē, holding it tightly so it couldn't get away. It was tangled in the netting, and with help she carefully extricated it. She remembered thinking how beautiful it was. She also remembered how overwhelmed she was, almost unable to believe that they had found the mythical *Notornis* alive and thriving.[3]

The next job was for Doc to take some photographs of what they were all sure was a nesting pair:

> Handling them gently, with due regard to their dangerous beaks, we took them to the beach, where they were securely tethered by one leg to be photographed, and then released

Joan Telfer, Neil McCrostie and Rex Watson with a pair of takahē on the shores of Lake Orbell.
ORBELL FAMILY COLLECTION

> to return as quickly as possible to the nest, which all the signs indicated was nearby, but which I decided must be left for another year rather than keep the hen too long away from the eggs, or run the risk of driving her away from the nest completely.
>
> A third bird was chased half-heartedly towards the bush, where in full view he fed and squawked for nearly an hour.

Everyone agreed the birds were striking. Doc wrote, 'the colouring was truly glorious – crayfish red beak, legs and feet; navy-blue head and breast, turquoise over the back, lower teal-blue, and finally tail-feathers in olive green'.

The team couldn't believe their luck. Although it was only 9.30 am, they ate their lunch. Then, as *Time* magazine later described it, they returned in a 'state of ornithological ecstasy' to Te Anau.[4] Geoffrey's daughter Lynley, who was nine years old at the time, recalled her father excitedly driving the boat round and round in victorious circles before coming ashore.

OPPOSITE: Geoffrey Orbell (left) and Neil McCrostie with one of a pair of takahē caught in the 1948 rediscovery of the species, in what is now known as Takahē Valley.
ORBELL FAMILY COLLECTION

Joan Watson (nee Telfer) always remembered how *Time* magazine described the rediscovery team as being in a 'state of ornithological ecstasy' when they found takahē in 1948.
WATSON FAMILY COLLECTION / ALISON BALLANCE

Once back in Te Anau, Doc got into his car straight away and drove to Invercargill to get the three reels of film and 12 photographs processed. Then the news broke. The *Southland Times* sent cadet reporter Bruce McNee to interview Doc, and although the local journalists didn't quite grasp the significance of the discovery, the story was soon making its way, via teleprinter, around New Zealand and the world. And the world, it turns out, was as excited about the rediscovery as Doc and the other members of the expedition. Within days, one of 'the greatest ornithological moments in history' was making headlines in newspapers and magazines,[5] and as the photographs winged their way to far-flung corners of the globe, the story was told and retold, earning the young cub reporter in Invercargill the nickname 'Scoop'.

Seventy years later, the McNee family would find itself caught up in the takahē story again, when Scoop's daughter Josie was working for construction firm Fulton Hogan, which had taken on sponsorship of the Takahē Recovery Programme. Josie was invited to help transfer and release takahē in Northwest Nelson – and among the birds being moved was one named Scoop, in honour of her father. But much would happen in takahē conservation before then, and before we fast-forward, we need to go back in time.

CHAPTER THREE

TAONGA

Takahē were first discovered by early Polynesian settlers, who arrived in Aotearoa around 750 years ago. In Te Waipounamu / South Island, the bird was called takahea, which means 'to stand up tall and stamp one's feet on the ground'. Northern Māori called the large Te Ika-a-Māui / North Island bird moho – it was known to be a solitary bird and, aptly, moho means 'hermit'. According to the online Te Ara – the Encyclopaedia of New Zealand, Māori described the takahē's night cry, a shrieked 'hu', as sounding like two pieces of pounamu being struck together underwater.[1]

Members of Te Waipounamu iwi Ngāi Tahu have a rangatiratanga, an obligation, to be kaitiaki or guardians for the takahē, which they hold in high regard as a taonga, a treasure. The iwi has a long and valued relationship with the takahē. Ngāi Tahu's online Kā Huru Manu cultural atlas records that 'Takahea were once plentiful on the shores of Moturau (Lake Manapōuri) and Te Ana-au, where they were gathered by hunting expeditions from Murihiku (Southland). The winter snow drove takahea down from the mountains to the lake shore, where they were gathered.'[2]

Bones found in middens across the country show that Māori ate takahē, which would have been a good-sized and succulent meal. Ngāi Tahu also traditionally used takahē bones to make needles and other items, and takahē feathers for weaving into kākahu cloaks. The feather colours are so vibrant that it's no wonder master weaver Erenora Puketapu-Hetet added some purple takahē feathers to the top decorative border of a modern kahu kiwi (a kākahu made primarily from kiwi feathers) that is ceremonially worn by dignitaries visiting the Museum of New Zealand Te Papa Tongarewa (Te Papa). These days, dead takahē whose bodies are in good condition are returned to Ngāi Tahu. Intact birds might be taxidermied so local marae can exhibit them in their whare or museum, to give people the opportunity to see the size and beauty of this taonga species, or takahē feathers can be made available for weaving.

But to Ngāi Tahu, takahē have a spiritual as well as practical significance. In the Māori world view, Te Ao Māori, taonga species are regarded as ancestors and are considered part of an iwi's whakapapa or genealogy. 'It is part of who we are. It's so much bigger than mahinga kai [food gathering], and being able to eat or wear them,' says Estelle Pera-Leask, former iwi representative on the Takahē Recovery Advisory Group. 'They are a fabulous bird. They represent New Zealand, and they represent what makes us unique as Māori. Without these species we have no culture, no reo or language, because our language is the language of the birds.'[3]

In the Ngāi Tahu Claims Settlement Act 1998, the Crown formally acknowledged the cultural, spiritual, historical and traditional association of Ngāi Tahu with more than 60 bird species – including takahē – as well as more than 50 plant species and six species of marine mammal. Under the Act, Ngāi Tahu is an equal partner with the Department of Conservation (DOC) when it comes to decisions about takahē, and the four southern Ngāi Tahu papatipu rūnaka – Ōraka Aparima, Waihōpai, Awarua and Hokonui – are a key part of the Takahē Recovery Programme. There is an iwi representative on the Takahē Recovery Advisory Group, a position held by Gail Thompson at the time of writing, and Ōraka Aparima Rūnaka is also actively involved in the takahē breeding programme as there are takahē living on an island within their rohe or territory.

Even though takahē have been moved around New Zealand, into the rohe of other iwi, the birds still belong to Ngāi Tahu, which is why the iwi takes an active role in translocations. It establishes whanaungatanga, relationships, with the mana whenua of the area to which takahē are being moved, and hands over the responsibility of kaitiakitanga for the taonga. 'It's part of our obligation to be protectors,' says Estelle.

Māori have a holistic perspective about conservation, which recognises that a threatened species such as takahē can be saved and have its mauri or life force restored only if the environment in which the birds live is secure and healthy. Estelle says Ngāi Tahu recognises the threat that introduced predators and browsers pose to takahē and other at-risk taonga species, and she believes the iwi has an important role to play in 'planting a pounamu wedge in the door of extinction, which is closing quickly on our [native] species'. She says the manu, the birds, need to thrive, not just survive. 'As an iwi we cannot do enough or speak openly enough about the need to hold the line for us and our children after us.'

Takahē place names[4]

The takahē's importance to southern Māori is reflected in the number of original place names in Te Rua o Te Moko / Fiordland which reference them.

One of Lake Orbell's original Māori names is Kōhaka Takahea, meaning 'nests of the takahē'. It is also known as Te Wai-o-Pani, meaning 'the lake of the friendless'.

Mauka is maunga or mountain in the dialect of Te Wai Pounamu, and according to Kā Huru Manu, the online Ngāi Tahu cultural atlas, Maukatakahea is the Māori name for Mt Philipps, located in the southern part of Mauīkatau / Resolution Island, in Tamatea / Dusky Sound.

Te Kāika-o-te-takahea, or 'home of the takahē', is a long, narrow inlet within the inner reaches of Taiporoporo / Charles Sound. It's also known as Emelius Arm.

Takitakahea, the 'cry of the takahē', is the Māori name for Neck Cove, a small bay on the eastern shore of Te Moeanu or Te Awa-o-Tū / Thompson Sound.

Camp Oven Creek on the Milford Track is Awatakahea, meaning 'takahē stream', and Forest Burn, in the Kepler Mountains, is Waitakahea, or 'takahē stream'.[5]

Slightly further afield and more recently named is Takahe Nunatak, a 1,000-metre-high rock peak close to Mt Bird in Antarctica; right next door is the slightly higher Kakapo Nunatak. And the Sign of the Takahe is one of several well-known neo-Gothic rest houses that were built on Christchurch's Port Hills in the early 1900s to serve as tea rooms for walkers. There are also plenty of 'Takahē' streets and other recent geographic uses of the name.

A search of the online New Zealand Gazetteer for place names including moho, the North Island takahē, comes up with a few hits. But while all are in Te Ika-a-Māui, none have any explanation for the likely source of the name, so may or may not relate to the bird.

CHAPTER FOUR

GIVE A BIRD A NAME

When the takahē was dramatically rediscovered in 1948, after 50 years with no official sightings, this just added to the general air of mystery and romance already surrounding a bird that was known to Western science only as four complete specimens and a single skeleton.

These days the South Island takahē, as it is formally recognised, is known to ornithologists as *Porphyrio hochstetteri*. But when its close relative the extinct North Island takahē was first described by the noted British anatomist and palaeontologist Richard Owen, he created a new genus for it: *Notornis*. Owen is best known for coining the name Dinosauria, meaning 'terrible lizard'. He is also famous for recognising the existence of giant flightless moa from a single fossil bone that was sent to London in 1839 to be examined. He went on to describe a number of moa species, beginning with the North Island giant moa (*Dinornis novaezealandiae*, literally 'the great or surprising bird of New Zealand').

In 1847, the naturalist and, later, politician Walter Mantell collected a number of large subfossil bones, including moa, from a midden on the Waingongoro River in Taranaki. It's important to note that these were just a few bones and they were collected in the North Island. Mantell was the son of another famous British palaeontologist, Dr Gideon Mantell, who first recognised and described dinosaurs (see box), and his father encouraged him to send the bones to Owen.

In 1848, Owen presented his findings on the bones to the Zoological Society of London. He certainly had a knack for drawing significant conclusions from a small number of fossils. From that North Island skull and a few other bones he recognised a large flightless bird that was similar to, but not the same as, a weka or pūkeko (which he called a purple coot). He suspected it was likely extinct. The genus name *Notornis* coined by Owen means 'southern bird' and comes from the Greek words *notos* ('south') and *ornis* ('bird').

Gideon Mantell's dinosaur fossils

When Gideon Mantell died in 1852, his son, Walter, inherited his collection of dinosaur fossils and had them shipped to New Zealand. In 1865, Walter Mantell gifted the best specimens to Wellington's Colonial Museum, including the tooth found by his mother, Mary Ann Mantell, from which Gideon Mantell described the *Iguanodon* in 1825. This was the first dinosaur to be described, and the tooth remains one of Te Papa's most valuable natural history treasures.

He named the species *mantelli*, in recognition of its discoverer, and the type specimen was lodged in the British Museum.

This was a newsworthy story – in the years that followed, newspapers as far-flung as the *Dundee, Perth and Cupar Advertiser* ran stories about the discovery and the strange birds to be found in New Zealand. Completely coincidentally, in 1848 – the same year Owen identified and named *Notornis*, and exactly a hundred years before the rediscovery of takahē in Fiordland – surveyor James McKerrow named a peak in the Murchison Mountains after him. Mt Owen lies to the west of the Mystery Burn, at the head of Sheerwall Creek.

In 1849, two years after the discovery of the bones, the first live takahē was collected, this time in the South Island, in Fiordland. Luckily for science, Walter Mantell was on hand again. According to a later account presented by his father to the Zoological Society in London, the takahē was found by sealers working in Tamatea / Dusky Sound.

> Perceiving the trail of a large and unknown bird on the snow with which the ground was then covered, they followed the footprints till they obtained a sight of the *Notornis*, which their dogs instantly pursued, and after a long chase caught alive in the gully of a sound behind Resolution Island. It ran with great speed, and upon being captured uttered loud screams, and fought and struggled violently; it was kept alive three or four days on board the schooner and then killed, and the body roasted and ate by the crew, each partaking of the dainty, which was declared to be delicious. The beak and legs were of a bright red colour.[1]

Walter Mantell managed to obtain the skin and bones of the bird, and labelling them '*Notornis mantelli*', he shipped the first complete takahē specimen off to his father in England. Historian John Hall-Jones writes that Gideon Mantell offered the preserved remains to the British Museum for £25, but this was rejected until he had paid for them to be stuffed and mounted. This first specimen of a South Island takahē remains in the Natural History Museum in London.

In 1851, Walter Mantell got hold of another takahē specimen, again in the southwest corner of the South Island. This one was collected by Māori whalers on Secretary Island, in Te-Awa-o-Tū / Thompson Sound, which runs into Doubtful Sound / Patea. It was duly sent to the British Museum but was returned to New Zealand's Dominion Museum (now Te Papa) in 1953 as a gift to the people of New Zealand following the species' rediscovery in 1948. This bird is the second-oldest New Zealand bird-skin specimen in Te Papa's collection.

Also in 1851, Captain John Stokes and his crew aboard HMS *Acheron*, accompanied by the schooner *Otago*, were in Fiordland as part of an extensive coastal survey of New Zealand. Hall-Jones records that while they were in Milford Sound / Piopiotahi, a note was made in the *Acheron* journal of 'a mysterious bird of "all colours", with an intense and magnificent plumage. As the journal contains a very complete list of birds, including all those likely to be confused with notornis, there is little doubt that Hansard [the journal writer] is giving us one of the earliest recordings of this rare species.'[2]

Despite these records, there was a general feeling of pessimism about the future of the takahē at the time. In his 1851 book *Fossils of the British Museum*, the older Mantell used an illustration of *Notornis* as the frontispiece, and wrote that 'it is unlikely that any further living specimens will ever be found'.[3] This was the first time that the takahē was effectively declared extinct.

Certainly, things were very quiet on the takahē front for almost 30 years. It wasn't until December 1879 that the third live bird was found, 14 kilometres southeast of

Takahē specimen collected by Walter Mantell in Te-Awa-o-Tū / Thompson Sound, Fiordland in 1851. After residing in the British Museum for 100 years, it was gifted to the Dominion Museum (now Te Papa) in 1953. TE PAPA

Lake Te Anau. This bird was run down by rabbiter Robert Scott's dog and taken back to camp, where the farm manager recognised it as a possible takahē. He carefully preserved the skin and bones, and the specimen was later bought by the Königlich Zoologisches und Anthropologisch-Ethnographisches Museum in Dresden, Germany. The museum director, Adolf Meyer, was creating a classification of all the museum's birds, including the new takahē specimen. On close inspection he decided that there were enough size differences between this skeleton and that of Owen's North Island bones for them to be considered separate species. In 1883 he published his work on bird skeletons, in which he named his South Island takahē *Notornis hochstetteri*, in honour of the Austrian geologist Ferdinand von Hochstetter.

Hochstetter had spent a productive year in New Zealand in 1859 travelling through the North Island and northwestern South Island to carry out geological mapping at the request of the provincial government in Auckland. He had returned to Europe and subsequently written more than 20 reports and scientific papers, as well as an influential book, *Neu-Seeland* (1863), on his travels.

The Dresden specimen remained on display until it was destroyed in the bombing raids that devastated much of the city during the Second World War.

Two more takahē specimens were collected in the late nineteenth century. In 1884 the partial skeleton of a takahē was found on the shores of Lake Te Anau. It was discovered by Richard Henry, later the custodian of Resolution Island in Tamatea / Dusky Sound but working at that time as a shepherd and farm hand. The following year he wrote to the Australian Museum in Sydney, explaining:

> I have an almost perfect skeleton of a Notornis mantelli which I want to sell by tender. I will advertise it & see what I can get for it. Please let me know what you would give for it. All the large bones are perfect and there are 13 vertebrae but most of the toes and 'small' bones of the wings, and ribs, are missing. I have been 5 years about these lakes as a rabbiter and bird hunter and the above is all I have seen of the takahe.[4]

The specimen was eventually bought by the Otago Museum.

In 1898, the fourth live specimen was collected by Jack and Donald Ross. They were camping on the shores of Lake Te Anau's Middle Fiord, not far from the mouth of the Snag Burn, when they heard an unusual bird call, which they noted was both double and deep. While they were out walking later that day, their dog went in pursuit and returned with a bird, which died shortly afterwards. Suspecting they had found a takahē, they dutifully rowed down the lake – in darkness – to Te Anau to dispatch it by mail coach to Invercargill early the next morning. A Dr Young telegraphed Professor William Benham at the University of Otago, and Mrs Young was sent up to Dunedin on the overnight train bearing the precious bird. Luckily, it was the middle of winter and temperatures were cold, so the bird was in a good condition for Professor Benham to dissect. The whole specimen – skin, bones and even its internal organs – was preserved and it was later mounted in the Otago Museum. This was the takahē that 20 or so years later caught the attention of the young Geoffrey Orbell.

For the following 50 years, from the discovery of that 1898 specimen until 1948, takahē were officially regarded as extinct. But in addition to the official records of takahē, it turns out that there were plenty of informal sightings and records, many of which Doc Orbell was aware. He collected all the sightings he could and plotted them on a map of Fiordland, which eventually gave him the idea to narrow his search to the Murchison Mountains.

Looking back, it is apparent that right up to the mid-twentieth century modest numbers of takahē survived across much of Fiordland. But this is a vast wilderness that attracted very few visitors, so it was easy for takahē to go largely unnoticed or at least unremarked upon.

John G. Keulemans' chromolithograph of 'Moho or takahe, Notornis mantelli' appeared in Walter Buller's 1888 book *A history of the birds of New Zealand*. **ALEXANDER TURNBULL LIBRARY REF: PUBL-0012-32**

J.G. KEULEMANS DELT & LITH. JUDD & Co LIMITED, IMP.

MOHO OR TAKAHE.

NOTORNIS MANTELLI.

(ONE-FIFTH NATURAL SIZE)

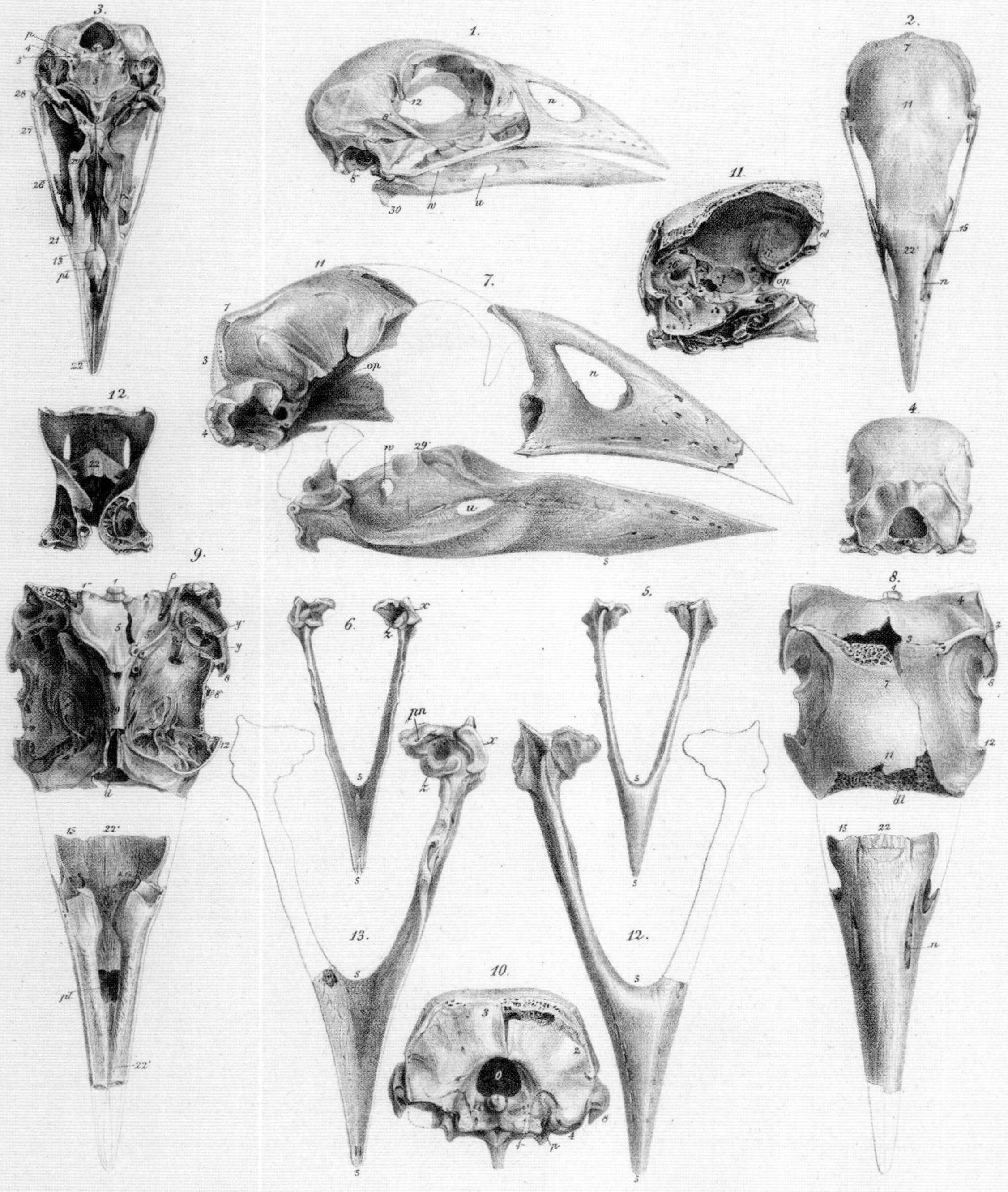

J. Dinkel del. on Zinc by J. Erxleben

Nat. Size.

Day & Son Lithrs to the Queen

1.–6. *Porphyrio.* 7.–13. *Notornis.*

Even great natural history observers and explorers such as Charlie Douglas in South Westland and Richard Henry in Fiordland spent decades in the late 1800s and early 1900s looking for takahē and failing to find them. In a study of New Zealand birds written in 1899, 'Mr Explorer' Douglas wrote:

> At one time [*Notornis*] no doubt flourished all over Westland, that is if we are to believe the Maories; but it is impossable [*sic*] it can now be in the country. I have been all over Westland from one end to the other, and always with a good bush dog, and if the bird existed, I could not possibly help finding it.[5]

He quite presciently predicted that 'if ever found it will be in the Lake and lagoon covered plateaus which are said to abound in some parts inland from the Sounds',[6] referring to the fiords. And there are indeed plenty of small lakes in the Murchisons, which lie not far as a takahē walks from Te Kāika-o-te-takahea, literally 'home of the takahē' and also known as Emelius Arm, in Taiporoporo / Charles Sound.

In late 1898, following the fourth takahē discovery, Richard Henry, who was by then working on Resolution Island, wrote a letter to his employer, Commissioner of Crown Lands James P. Maitland. He described a trip he made into the alpine zone above Tamatea / Dusky Sound in search of the elusive takahē. Henry prefaced the letter with this:

> I had the idea that the takahes were all dead because they must have been rapidly decreasing, or how could the Maoris become acquainted with them if the birds were anything like as scarce as they are in our time. Therefore I had no faith in my hunt till the finding of this last one gave me reasonable hopes. I several times had an idea that there was a stranger [unknown bird] about but always gave it up because I could never hear a strange voice night or day though I could hear all the others.[7]

Henry was wondering if takahē might be nocturnal, which might explain why he never heard them during the day. However, after spending a few nights camping up high, he wrote that there were 'roas [kiwi] calling long before dark & after daylight in the morning but no takahes. They are either dumb or don't live up there though it is just above where I suspected strangers.'[8]

In another letter, written in early 1899, Henry pondered how he might catch a takahē if he came across one.[9] He had a trained bird dog, but thought that herding the birds into stretched fishing nets would work; it was an idea that Orbell later put to good use.

A comparison of pūkeko and North Island takahē bones, from Richard Owen's *Memoirs on the Extinct Wingless Birds of New Zealand*, 1879. UNIVERSITY OF TEXAS LIBRARIES

Despite his efforts, Henry never did come across a live takahē, and he laid the blame for their apparent demise at the feet of introduced predators such as rats. But the lack of sightings didn't stop him speculating on the takahē and its likely evolution:

> The takahe is only a little way removed from the swamp hen, to adapt it for living in the bush. And without doubt its forefathers came here on the wing, but, becoming tired or storm-bound, dropped down in the West Coast bush, where, before the advent of rats, they found no enemies on the ground, and abundance of good seeds and other food. Then, if there were a few bush-hawks about these well-fed hens would not care to fly until they were too fat to do so. On the wing swamp hens are very easy game for the hawk, for I have seen him knock down several. And if some of those fat ones tried to get away and escaped a knock or two from the hawk they would be content to remain in the bush. Then with mates of like experience they could found that notable branch of the family that we call *Notornis*.[10]

The changing name

For the first few decades following Richard Owen's description of the takahē, people referred to it as *Notornis mantelli* – even after 1883, when Adolf Meyer described a South Island specimen as a different species, *Notornis hochstetteri*. Fiordland naturalist Richard Henry, for example, wrote in the *Transactions of the New Zealand Institute* in 1899 'On the Probable Origin of *Notornis mantelli* and its Extinction in New Zealand'.

Eventually, the name *Notornis hochstetteri* began to be used more commonly, and it was this name Geoffrey Orbell and others were using in 1948, often shortening it to 'the *Notornis*'. But just a few months after the rediscovery, prominent ornithologist Robert Falla, writing in the journal *Emu* about the first scientific trip into Takahē Valley in January 1949, was already casting doubt on the validity of separating the North and South Island birds. He commented that he thought Meyer's description varied from Owen's original one by just a 'few differences of doubtful significance'.[11] It turned out that other taxonomists shared his doubt, and over the next few decades the North Island and South Island takahē were relegated to subspecies of *Notornis mantelli*.

Taxonomists also decided that the birds belonged to *Porphyrio*, the same genus as the pūkeko, and the genus name *Notornis* was dropped. The North Island birds became known as *Porphyrio mantelli mantelli*, while the South Island takahē were *Porphyrio mantelli hochstetteri*.

In 1996, evolutionary biologist Steven Trewick concluded that Meyer had been right all along, and that, based on the morphology of bones and on genetics, the two were indeed separate species. The original species names were reinstated, giving us *Porphyrio mantelli* for the extinct moho or North Island takahē, and *Porphyrio hochstetteri* for the South Island takahē.

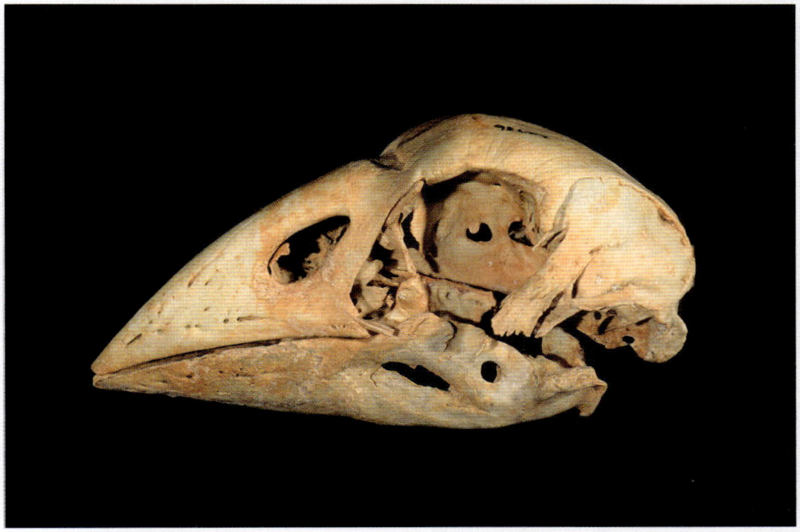

The moho or North Island takahē was described by Richard Owen from subfossil bones collected in Taranaki. **ROD MORRIS**

CHAPTER FIVE

FROM DREAM TO REALITY

We carry our backpacks and a box of food on a well-trodden path through the tussocks and hebe bushes to the historic Takahē Valley Hut. The hut has a coat of slightly peeling, dark green paint and flaunts a large solar panel, but in every other respect it is a basic corrugated-iron affair that sits squarely against the edge of the sombre forest, much as it has for nearly 70 years.

A large chimney made from rusty flat iron partly obscures a window and sits oddly off-centre from the hut's original rectangular roof gable. A large lean-to on the hut corner is a later addition; it shelters the battered front door, whose hinges squeal in protest as we push it open. The hut is small and sparse: an uneven wooden floor, four bunks, a kitchen bench, an open fireplace. The two small windows are covered in spider webs, and there are gaps aplenty. There is a gas cooker, a motley assortment of battered pots, crockery and cutlery, and the usual backcountry collection of old condiments. It's not much, but it's all the shelter you need in the hills.

Takahē Valley Hut was built in 1954 and is the oldest hut in the Murchison Mountains. For the first five years following the takahē rediscovery, visiting parties camped where the hut now sits. Finally – after much urging, which Doc Orbell began as early as 1949 – the Fiordland National Park Board agreed that the Wildlife Branch of the Department of Internal Affairs could build a small hut. It was a boon to the deer hunters as well as the takahē researchers.

The contract to build the hut went to Amphibious Airways Company, based in Invercargill, and the timber framing and corrugated-iron cladding were delivered to the valley by flying boat, which landed on the lake. These days the hut is lined, although that improvement didn't come about until the mid-1980s. But for the earliest researchers, there was no option but to sleep under canvas in rudimentary camps.

****

Takahē Valley Hut is the oldest hut in the Murchison Mountains. WILLIAM LEE

Hot on the heels of the November 1948 rediscovery, New Zealand's leading ornithologists began beating a path to the Tunnel Burn – soon to become known as Takahē Valley. They were keen to study the mysterious bird, about which nothing was known, and learn as much as they could of its biology. In January 1949, the first party arrived. At the request of the Department of Internal Affairs, which at the time was responsible for the flora and fauna of New Zealand and its preservation, Doc Orbell was guide and leader. The advance party of Doc, Internal Affairs field staffer Frank Woodrow and Canterbury biologist Baughan Wisely arrived first, to establish a suitable campsite near the outlet of the lake, where Takahē Valley Hut would later be sited. Later, in the *Illustrated London News*, Doc wrote:

> The tents having been erected. I set out with Mr Wisely to reconnoitre the lower end of the valley. Notornis signs were everywhere in evidence, and very soon we heard the characteristic 'thumping' sound and located a pair of Takahe within 50 yards of the spot where we had made our capture early in November. These birds were not disturbed, and we returned to camp with the knowledge that there seemed to be at least one pair for observation.[1]

The next day the rest of the party arrived. In addition to Joan Telfer, Neil McCrostie and Rex Watson from the original discovery party, there was Robert (Bob) Falla, director of

Incubating takahē, seen in December 1949 during the fourth Takahē Valley Investigating Party expedition. E.P. CHRISTENSEN, ARCHIVES NZ - COMMUNICATE NZ COLLECTION R24457518

the Dominion Museum and one of the country's leading ornithologists, and botanist and ornithologist John (Jack) Sorenson. Doc wasted no time in taking Bob out to show him a takahē. The discoveries came thick and fast.

> The most interesting discovery that afternoon was the finding of a nest in which were the remnants of an egg, and the dead, and somewhat decayed, body of a chick. The nest was made up of soft stalks of grass laid to a depth of at least 3 inches between two very large snow grass clumps which bent over it, giving a back and front entrance and also perfect shelter from the weather … The shell was in many small pieces, which were carefully collected so that they could be fitted together to make at least a portion of the only Notornis egg then in existence. The chick itself appeared to have had the back of its skull crushed in.

The dead chick and the eggshell provided two important clues. The first was that takahē lay two eggs, and the second that stoats were a threat to the birds. Doc wrote, 'it seems probable that a stoat found the egg unprotected for a moment, and having broken the shell, was about to devour the chick when the parents returned … These pests are now all through the wildest bush areas.'

The ornithologists went on to have a very successful trip, collecting important basic information on the biology of the takahē. On day two, they spread out in a line across the valley to search for birds and nests. Jack soon found a bird sitting on a nest, and as the adult left by the back entrance a small chick ran out into the waiting hands of Doc. Photos and measurements were taken of the fluffy black youngster and its parent. The latter hung about anxiously nearby, calling to the chick, which squeaked in reply until the two were reunited, disappearing into the thick tussock.

The party came across plenty more empty nests. Then just before lunch Doc

> parted two large clumps of snow-grass with my hands and found myself looking at a sitting Notornis from a distance of 1 ft. The surprise was no doubt mutual, but the bird recovered first and made a hurried exit, leaving in the nest one large egg. Truly a joy to the eye – dull cream in colour, with brown spots and faint mauve blotches. That the egg was fertile could be proved by listening to the movements of the embryo chick getting ready to open his door to the world.

The egg was photographed and returned to the nest, and the team moved on to allow the bird to return to incubation.

Zoologist Louis Gurr takes notes as Robert Falla measures a takahē egg, December 1949.
E.P. CHRISTENSEN, ARCHIVES NZ - COMMUNICATE NZ COLLECTION R24457502

The team members counted 14 different adults that day, and noticed that many of them were making contact calls with unseen chicks. Doc commented how quickly the birds, which were very tame on the first day, became wary of people.

There were lots of useful observations about the bird's biology. Bob paid careful attention as he watched a takahē feeding on snow tussocks, stripping seedheads from the flowering grasses and plucking individual stalks or tillers to snip off only the tenderest base before discarding the stalks in a growing pile. Meanwhile, Doc watched two birds make loud contact calls across the lake, before one bird began to wade towards the other. He commented that the bird didn't mind getting its lower body wet.

Over the following days the party searched the whole valley thoroughly, including the high slopes and basins. They discovered more birds in the nearby Point Burn Valley, which was easily reached from Takahē Valley. Based on the number of nests found, Bob estimated there were at least a hundred birds, but sightings on subsequent trips later the same year and the discovery that a single takahē pair makes a few trial nests before settling on one, saw that population estimate significantly downgraded to about 20 birds in both the Point Burn and Takahē Valleys.

The next scientific expedition headed into Takahē Valley in late winter. Braving the snow and cold were Graham Turbott from Auckland War Memorial Museum; scientist Gordon Williams, who had just been appointed to the Department of Internal Affairs; and field officer Ken Miers, who was employed at that time as a deer-culler. For Gordon, it was the beginning of a decade-long association with the birds.

A few days before their trip, Doc had been into the valley and reported that the lake was frozen solid and there was snow and ice lying on the valley floor. He saw two takahē in the valley, and a further five in the Point Burn. A few days after this visit, Ken visited the valley to organise the field supplies, which had been dropped in by air, and he commented on a recent dumping of fresh snow.

Bad weather hampered the expedition. Heavy rain alternated with snow to keep the team tent-bound for much of the time, but they ventured out during breaks in the weather. On the first clear afternoon, they were walking along the northern shore of the lake when they caught a glimpse of a takahē, which fled at the sight of them. 'It kicked up a spurt of snow as it plunged beneath the fringing scrub.'[2] They found feeding sign, droppings of various sizes and footprints, and heard individually recognisable calls. From this they estimated there were five birds around the lake, but they didn't manage to see any of them.

The trio made it through deep snow on the ridge to the south to visit the Point Burn, which is at a lower altitude and had less snow lying on the ground. On the grassy flats they saw one takahē, which tolerated them for a while before finally breaking into a run and disappearing into the forest.

The trip was frustrating for the scientists, but it did add a vital piece to the picture

The fourth Takahē Valley Investigating Party photographed at their forest edge camp in December 1949. Left to right standing: ornithologist Robert Falla, field officer Ken Miers, movie and still photographers D. Oakley and E.P. Christensen. Sitting: zoologist Louis Gurr, J. Ollinshaw and a packman. E.P. CHRISTENSEN, ARCHIVES NZ - COMMUNICATE NZ COLLECTION R24457491

they were beginning to build of this little-known bird. They concluded that heavy snow covering the snow tussock for long periods would make it difficult for birds to find food throughout winter, and they suspected that at least some of the population must move away for at least some of the time, probably down into the forest. They commented that the droppings of the bird they saw in the Point Burn, which was eating snow tussock, had the same chaff-like consistency as the droppings seen in summer, whereas droppings they collected in Takahē Valley were dark green in colour and of a soft texture, suggesting the birds were eating something different. This movement between the subalpine zone and the forest tied in with earlier suggestions that the four sightings of takahē in the late 1800s were of birds near the lakeshore or seashore, driven to low altitudes by cold winter weather.

The next summer was a busy one, with three expeditions organised to follow a complete takahē breeding season. Frank Woodrow, Gordon Williams and Ken Miers headed in for a month from mid-October 1949, to try to catch egg-laying and incubation. They mapped the territories of five pairs in the valley, which appeared to be the same pairs seen the previous year. On 25 October they found a bird sitting on two eggs, and for the next few weeks they observed as closely as they dared without disturbing it. The scientists had elected not to mark the adults, so they couldn't tell them apart, which meant they weren't sure whether both parents took turns incubating or whether just one bird had this role.

When Bob Falla and the next party arrived in the valley on 1 December 1949, the same pair were still incubating. There was a big appetite from the public for news of this celebrity species, and on this trip photographers from the National Publicity Studios and the National Film Unit joined the group. The resulting newsreel of the expedition is notable for how tolerant and unafraid of humans a nesting takahē was in those early years. Staying on the eggs, the nesting bird in the film allows Bob to stroke its rump, and then it has to be coaxed off its eggs. It walks close by the camera operator, and returns as soon as it can to the nest, undeterred by the camera team filming at close quarters.

The nest was very well photographed, before a nest check 11 days later finally revealed that the eggs were well decomposed. As a result, the scientists couldn't work out how long the clutch of two eggs would normally be incubated for.

Other takahē pairs were having better luck. Soon after Bob and his party arrived in the valley, they observed one pair with a small chick, just two or three days old, while another pair had a youngster that was already two to three weeks old. The remains of invertebrates in the youngest bird's droppings revealed it was being fed almost entirely on insects, including many aquatic insects and larvae collected on the lakeshore. After several weeks the chick's diet changed almost completely to the same tussock bases its parents were feeding on.

Gordon Williams and the eminent scientist Charles Fleming joined the third party, which made a short trip into the valley in early February 1950. Of the eight or nine takahē pairs that were being monitored in both Takahē Valley and the Point Burn, only one pair still seemed to have a half-grown chick. Another dead fledgling was found, and there was no sign of any other chicks.

The scientists were now confident that although takahē laid two eggs, they raised only one chick, but they weren't sure of the reasons for the poor breeding success. They speculated about low fertility and predation by stoats, and worried that such a low rate of productivity didn't bode well for the small population.

However, they were also heartened by news that the takahē appeared to be more widespread than first thought. Expeditions by Department of Internal Affairs staff, especially field officer Frank Woodrow, had confirmed the presence of birds in seven valleys in the Murchison Mountains, and there were still other valleys to check.

In 1949, soon after the rediscovery, the government passed legislation to create a 530-square-kilometre 'special area' to protect the takahē. The area covered the Murchison Mountains, and entry was prohibited except to bearers of a permit from the Commissioner of Crown Lands. In addition to surveying for new takahē areas, Frank began trapping stoats and controlling deer and possum numbers. He also acted as an on-the-ground guard to

The last *Notornis*

The only *Notornis* remaining is the journal of Birds New Zealand (formally known as the Ornithological Society of New Zealand). The society changed the name of its bulletin in 1950, from *New Zealand Bird Notes* to *Notornis*, in recognition of the significance of the bird's rediscovery. The first volume under the new name, published in July 1951, included a series of scientific papers on its namesake by leading ornithologists of the day, and these were bundled together and published for the general public under the title *The Takahē*. The takahē remains the logo of the society, and a fibreglass takahē model named Orbell has been the society's mascot for the past 30 years or so.

ensure there were no unauthorised visitors. The local park ranger, meanwhile, had the job of controlling tourists visiting the glow-worm caves, to make sure no one sneaked away to visit the takahē.

Three years later, in 1952, Fiordland National Park was finally created, covering the vast area that had been set aside as national reserve since 1904 (the first iteration of the park excluded the Hollyford Valley, which was added later). The Takahē Special Area remains in place today.

Shortly before the rediscovery, the Native Bird Preservation Committee (later called the Fauna Protection Advisory Council) had been set up to advise the Wildlife Branch on how it should manage threatened species, and this external committee went on to give advice on takahē policy and operations until 1987. At that point the New Zealand Wildlife Service, as the Wildlife Branch was later known, became integrated into DOC.

These early efforts in takahē research and conservation management were pioneering. They heralded what would become a hallmark of New Zealand conservation: innovative and successful threatened species management. Having been going for 75 years, the Takahē Recovery Programme is now the longest-running bird conservation programme in New Zealand, and one of the longest in the world.

Shaped by ice

Takahē Valley is a secret fortress, a lost world ringed by rugged peaks and steep bluffs. It lies on the eastern flank of New Zealand's greatest wilderness – Fiordland. This is a landscape sculpted by ice. To the west lie 14 mighty fiords, carved by enormous glaciers that once flowed from the mountains well beyond today's coastline. These great glaciers have come and gone often over the past 2.5 million years, advancing seven times in the last 250,000 years alone. Time and time again, ice ages have held the land in a frozen grip, locking up so much water as ice that the sea level dropped as much as 120 metres lower than it is today.

Some 15,000–20,000 years ago, at the height of the most recent ice age (known as the Ōtira Glaciation), the southern glaciers were up to 1,400 metres thick. The glaciers coalesced, until all of Fiordland and the Southern Alps / Kā Tiritiri o te Moana were encased in ice and just the tops of the highest mountains could be seen above the frozen shroud.

These great ice streams poured west and east of the main divide, the rocks and debris at their base slowly grinding their way down through the tough, ancient bedrock below. The snouts of the glaciers were like giant bulldozers, shovelling mounds of rocks as they nosed towards the coast. In the west, the glaciers culminated at the much-reduced ocean, where their steep, frozen cliffs spawned icebergs. To the east, the glaciers spread out across the land to form a vast ice sheet.

During the interglacial periods, the ocean rose and flooded back up into the U-shaped glacial valleys in the west, resulting in dramatic fiords with distinctive steep sides and rounded bottoms. To the east of the main divide, water filled glacial troughs to form lakes such as Manapouri and Te Anau, the floors of which are now so deep that they lie below sea level. Great heaps of shovelled rock were left behind as moraines, marking where the ice had been. In higher alpine valleys, the last remnants of the melting ice created small lakes. Carved out of the sides of the valley are bowl-shaped cirques or hanging valleys, and the ridges are steep and sharp.

As the ice ebbed and flowed, so did the forests and the low alpine grasslands. Finding home in the grasslands were species such as the takahē, which arrived here about the same time as the first of the great glaciations. As temperatures chilled and the ice advanced again, forests retreated seawards and contracted in the warmer lowlands, while the grasslands expanded down the mountain slopes behind them. The process took tens of thousands of years, and as the tussocks slowly moved downslope, seeding into ground recently vacated by dying trees, the birds moved with it. They thrived in this changed world, with its vast swathes of grassland bordered by frequent patches of forest and shrubland.

When temperatures warmed and the ice tide turned, it was the forests turn to conquer.

Fiordland has been shaped by glaciers, which have resulted in distinctive u-shaped valleys, sharp ridges and small alpine lakes. Lake Creek, Murchison Mountains. **ANDREW DIGBY**

The trees chased the grasslands as they retreated upslope, and by the time the first Polynesian voyagers set foot in Aotearoa around 750 years ago, 80 per cent of the land was covered in forest. The diminishing glaciers left behind bare ground, soil and rock, providing a welcome substrate for the alpine vegetation as it withdrew to the high valleys and basins. The takahē world shrank in scale, from an empire spanning the land from mountains to sea, to smaller areas of grassland and shrubby edge habitat along the length of the South Island.

This pattern of glaciations repeated around 20 times as the ice ventured forth from its alpine lairs and then slunk back. The last glaciation finally ended around 11,000 years ago, and as global temperatures warmed, the ice began its last great retreat. Since then, temperatures in the current interglacial have oscillated and glaciers have had small surges and retreats, but nothing on the scale of the ice rivers that shaped the dramatic landscapes of the South Island. And in a warming world, it feels unlikely we'll ever see so much ice again.

Māori tell a different story of how Te Rua o Te Moko / Fiordland was created. The region represents the raised-up sides of *Te Waka o Aoraki*, the canoe of Aoraki, after it foundered on a submerged reef. Its occupants, Aoraki and his brothers, were turned to stone, becoming the highest peaks of Kā Tiritiri o te Moana, the Southern Alps. Atua or demi-god Tū Te Rakiwhānoa carved the fiords out of the raised side of the wrecked *Te Waka o Aoraki* in an effort to make the region habitable by humans. The deep gouges and long waterways of the fiords were created to provide safe havens along the rugged coast, and were stocked with fish, forest and birds to sustain life.

CHAPTER SIX

THE HARD WORK BEGINS

The helicopter drops us on a high ridge, with Takahē Valley on our south side and the Ettrick Burn to the north. In the distance, Te Anau is still obscured under thick early-morning lake fog, but up here we are bathed in sunshine. Yuki the takahē tracking dog, an 11-year-old wirehaired pointer, is wearing her bright Conservation Dog Programme coat and a muzzle, which is currently full of snow as she has just dug her face into one of the late-season snowdrifts. Rangers Nichollette Brown and Glen Greaves are getting ready for day 1 of the 2017 takahē head count, a big job that takes place every three years. Other rangers have been dropped off at different sites across the Murchison Mountains, and over the next two late-November weeks, weather permitting, they'll visit every takahē territory to see which birds are in residence. The last count was in 2014, when 79 birds were counted.

We'll be passing through a number of takahē territories today, looking for fresh feeding sign and using Yuki's sharp nose to try to find birds and see if they're nesting. About 40 of the Murchison's expected 100-plus birds are wearing radio transmitters, and Nichollette is carrying a receiver and an aerial so she can scan for likely signals. The first north-facing valley we'll be searching is usually home to six takahē, but several territories have been empty since the 2007 stoat plague took out many birds. We spread out down the slope and then make our way across it, weaving through waist-high tussocks and scrub to see if any birds are in the area and if we can identify them.

Glen notices some feeding sign that is several weeks old, so while Nichollette begins scanning for likely birds, I turn on the FoxPro I'm carrying. It's a compact weatherproof loudspeaker, and I hit play and hold it above my head so the recording can carry as far as possible. The strident calls of a territorial takahē echo across the valley. The calls end and we all strain our ears in the quiet that follows, hoping for a reply that will give away the presence of a bird. Glen keeps an eye on Yuki, trusting that her sensitive nose will catch any scent of a bird on the breeze. Nothing.

Rangers search for takahē on a ridge overlooking the Ettrick Burn, Murchison Mountains.
ALISON BALLANCE

But then Nichollette shouts to say she has a signal. It's a female that was puppet reared in captivity at Burwood and released here several years earlier. The signal is coming from far down on the forest edge. Glen surmises that she has probably paired up, and that the pair have moved down to the breeding core of their territory and are sitting tight on a nest, which is why they are not responding to the calls. We decide it's too far to drop down for a closer look, but Glen is pleased that the bird seems to have found a mate, as last time he came across her she was on her own.

We sidle some more and try again. This time we hear two birds reply. Nichollette scans through the radio frequencies of a few takahē until she hears the telltale beeps for Te Maka. This seven-year-old female was also captive reared at Burwood. She was released here a few years earlier and is clearly with a mate too. Again, the birds are too far away to be worth the effort of a closer look, but Glen comments that he's pleased to see the territory is occupied and he makes a note on where we've found the pair.

****

Nichollette Brown, Glen Greaves and takahē dog Yuki near the head of Takahē Valley.
ALISON BALLANCE

In the early 1950s, Frank Woodrow was a very busy man. The experienced Canadian trapper, who had ostensibly 'retired' to New Zealand, was the man in the mountains when it came to takahē, and to trapping stoats and shooting deer. Between October 1951 and April 1952, for example, he made seven trips into the Murchison Mountains, spending a total of 97 days in the hills. For a month of that he was on his own. In the absence of huts, he camped or bivvied out.

By the end of those trips, Frank reported that he had shot 38 red deer, and trapped 19 stoats and 54 possums. A year later, the deer tally was reported to be 72 animals, 58 of which were shot in Takahē Valley and those neighbouring it.

Despite the high deer tally, which was the cause of some concern, Frank and the various people who accompanied him – especially his frequent companion Mr Morgan – were primarily trying to find out how many takahē there were and where. The birds were known to be in Takahē Valley, the Point Burn and the Mystery Burn, and Frank had already expanded their known range to a number of other valleys.

These trips were no small undertakings for Frank. While the eastern valleys in the Murchison Mountains such as Takahē Valley and the Point Burn provide easy travel, the country to the west is steeper, rockier and more difficult to traverse. A member of the 1955 Canterbury Museum Murchison Mountains expedition, which came in search of takahē

The recently rediscovered takahē featured on one of three stamps issued by the New Zealand Post Office to mark the Centennial of Southland in 1956. SHUTTERSTOCK

and also uranium (nothing of significance was found), later wrote in the *New Zealand Geographer*: 'The traveller is constantly skirting bogs, banks and bluffs and climbing up and down irregularities of terrain, with many detours made for tangles of trees and bushes within the forest itself. Put simply it takes longer to traverse country in Fiordland.'[1]

In the meantime, Department of Internal Affairs Wildlife Branch staff continued to make regular trips into Takahē Valley and the Point Burn to monitor the breeding success of the various pairs, whose territories were by now quite well mapped. The Wildlife Branch's science unit was a very small affair. Although it was tasked with leading research efforts into rare birds, and takahē and kākāpō in particular, a growing focus that came to involve much of the unit's effort was studying and improving populations of introduced

Bird banding in New Zealand

The early 1950s were early days for bird banding, both in New Zealand and overseas. In the first half of the twentieth century people often made their own bird bands as required, cutting up cigarette tins or aluminium pots, stamping them or scratching on numbers, and sometimes marking them with enamel paint. Different biologists marked different return addresses, such as 'Return Otago Univ' or 'Tell Times-Age'. Coloured plastic rings such as those used by poultry farmers and caged-bird enthusiasts were also sometimes used.

In 1947, the Department of Internal Affairs began its own scheme for banding gamebirds such as pheasants. In 1950, the Ornithological Society of New Zealand introduced a national banding scheme for all other kinds of birds, supplying aluminium bands in different sizes and keeping records of which birds were banded. This scheme was taken over by the Dominion Museum in 1962. The two schemes amalgamated in 1967 to become the New Zealand National Banding Scheme, run by the Wildlife Service and its successor, DOC.

All managed takahē are identified by a uniquely numbered metal band.
BRENDON DORAN

gamebirds, including ducks and pheasants. Science unit staff included Gordon Williams, who was there through the 1950s and took the lead with takahē, and ex-deer-culler Ken Miers, who had by this time been to university to study biology and who came back to work full-time in the Wildlife Branch from 1953 until 1966. In 1949, Ken had notably discovered an overhang above Takahē Valley containing evidence of its use as a shelter by Māori, including a woven sandal and bones of moa and takahē.[2]

By the end of 1952, there had been 10 expeditions into the Murchison Mountains, totalling 345 person-days in the field. But from the moment of rediscovery there were differing opinions about what approach to take: a cautious conservative approach or bold experimentation. The Forest and Bird Protection Society, for example, wanted to leave the birds undisturbed in their Fiordland hideaway. The declaration of the Takahē Special Area reinforced the idea that takahē were best left alone. Some people – Doc Orbell among them – already felt that the amount of research being carried out was disturbing the birds unnecessarily. Other people suggested that more intensive management was needed, such as captive breeding or translocation to an island. As part of this, people were initially reluctant to catch and band the takahē, even though marking individual birds is the only way to follow their breeding and survival, allowing much better estimations of how many birds there are. It was the beginning of a long-running debate about the best way to manage this threatened species.

Looking back on the first four years following the rediscovery, Brian Reid – who worked on takahē for the Wildlife Service in the 1960s and early 1970s – wrote in 1967 that: 'Considerable information resulted [and seven papers published] but the returns did not compensate for the time involved, as there was no way of identifying individual birds. This precluded reliable work on many aspects of social behaviour and population ecology.'[3]

Scientists had been urging the Wildlife Branch to band birds since 1949, and finally in late 1952, Ken Miers and Thane Riney banded the first takahē. In December, they caught nine birds – seven adults and two chicks in Takahē Valley – and fitted each with a different combination of coloured plastic bands on their legs. These bands were commonly used to mark poultry, and no one knew how well they would fare in a harsh mountain environment. A year later, a further nine birds were fitted with both plastic and numbered aluminium bands. The earliest plastic bands were not a great success: three were known to have been lost in the first year, some colours faded very quickly and another bird pulled its plastic band off as soon as it was released. But aluminium bands and plastic bands coated with varnish to keep them shut and to prevent fading showed promise, and tests by paint experts in the Dominion Laboratory at the Department of Scientific and Industrial Research (DSIR) helped identify exactly which colours faded least in the field. The biologists cautiously agreed to keep using them unless they were causing distress or harm to the birds.

The unique colour combinations allowed the observers to identify individual birds, but it still wasn't without its challenges as the birds' legs were often hidden by vegetation. In this case, there was nothing for it but to walk slowly towards the birds and encourage them to 'drift' into a clearer area where their legs could be better seen.

The bands soon showed their worth, allowing the biologists to collect data that couldn't be gathered when birds were unmarked. After two different birds were seen on the same nest, they confirmed that both sexes incubate. Intriguingly, they also noted that a one-year-old bird was observed incubating an egg, and that one bird was seen with two small chicks two months apart, raising the possibility that takahē nest twice. The former was more likely a helper bird, sharing incubation with its parents, and in the latter case it's likely that the first chick died early enough in the season for the adults to re-nest, although takahē generally nest only once.

Every year, scientists made a number of trips into Takahē Valley and Point Burn. Ken and Gordon were the main observers, but they were joined by other experts such as geologists, botanists and soil scientists to help build a better understanding of the environment and how the birds used it. Much of the research consisted of general natural history observations, and the hallmark of this decade is that these comments were faithfully summarised each year and published in the journal *Notornis* for the New Zealand ornithological community to ponder over and enjoy.

As the 1950s progressed, Ken found himself working more and more on gamebirds, especially swans, but Gordon continued to organise regular visits by Wildlife Branch staff and other biologists into Takahē Valley and the Point Burn to keep building a picture of the species' life cycle, its behaviour and, importantly, its breeding successes, death rate and overall population status.

Gordon's magnum opus, coinciding with his departure from the Wildlife Branch and summarising a decade of combined research since the rediscovery, was presented to the Canterbury branch of the Royal Society of New Zealand in 1959 and published in 1960 as a 23-page article in *Notornis*. In it, Gordon listed in great detail what had been discovered about takahē, based on the population of around 30 birds that had been observed in the two main eastern valleys. Overall, he suggested, the total population of takahē might be only a few hundred birds.

In 1960, the pace of visits to Takahē Valley slowed down. A policy of 'minimum disturbance' was implemented and a decision made to leave the population alone during spring. Banding of adults stopped for several years, although chicks were still banded. The first flurry of research had come to an end, and the next decade would be very different indeed.

Where takahē once were

Judging by fossil bones, takahē were once quite widespread across drier areas of the South Island, although it seems they were never abundant, which is not altogether surprising for a primarily grassland bird living in a mostly forested land. Fossil takahē bones have been recorded at more than 50 sites, from Nelson right down the east coast of the South Island and into Southland, but they are absent from high-rainfall areas such as the West Coast and Northwest Nelson. They have been found in dunes, caves and swamps, and in a few Māori middens. Almost all of the records are from lowland sites, often near the coast, which would have been a mosaic of forest, shrubland and grassland during the recent Holocene period, which dates from about 11,600 years ago, following the last major glaciation.

In 1974, Brian Reid published a paper listing all the records and sightings – both confirmed and unofficial – of takahē prior to their rediscovery in 1948.[4] There was a handful of reliable sightings in the Nelson district, and in Fiordland the records spanned from Martins Bay in the north to Lake Hauroko in the south, and even to the east of Te Anau as far as Mossburn. Brian suggested that by the mid- to late 1800s takahē were still spread 'irregularly and discontinuously' across Fiordland, and up until the 1940s there were reports from explorers and deer shooters of pūkeko (sometimes referred to as 'giant pukakis') in places that were much more likely to be home to takahē.

By the 1960s, however, the Murchison Mountains appeared to be the last remaining stronghold of the species. Wandering birds were, and even still are, occasionally recorded some distance away – for example, on the edge of the Kepler Mountains to the south, in the Stuart and Franklin Mountains to the immediate north, and even down at the control gates where Lake Te Anau exits into the Waiau River.

CHAPTER SEVEN

A CLOAK OF SECRECY

We climb back up towards the ridgeline and head east until we reach the next territory, which has also been empty for a few years. This time a single bird replies to the recorded calls. Then I see that the dog Yuki is indicating she has smelt something and we drop down a couple of hundred metres as she follows the scent trail. Yuki soon focuses on a large clump of tussocks. I'm lagging behind, and by the time I catch up I find Glen Greaves and Nichollette Brown looking at a large pile of fresh takahē droppings. It's a latrine, which means there's likely a nest hidden under the nearest tussock. Glen says they saw a takahē further down by the forest edge, which was the one calling, and also another bird. Judging from Yuki's intense posture, however, there is a third bird still incubating, so it seems this is a breeding pair with a juvenile helper from the previous year. None of the birds has a radio transmitter. Nichollette circles the dense clump, parting long, drooping stems until she finds an obscured tunnel leading to the interior. With her head torch on, she kneels and peers inside. 'Two eggs,' she announces. 'Very warm.' The incubating bird has only just vacated and, unfortunately, we didn't manage to get a glimpse of the colour of its leg bands to identify it.

Glen quickly opens his pack and gets out the equipment he'll need to check the eggs' fertility. This includes a thick black cotton bag, a candling torch with a dark egg-shaped cup of foam around the bulb end, and a sheet of diagrams detailing what egg development – veins, air-sac size and embryo size – he can expect for each day of the 30-day incubation period.

Holding the candling torch in one hand, Glen slips the cloth bag over his head, inserts his hands through the short sleeves on the sides and then pushes one hand out through a slit along the bottom of the bag. Nichollette carefully places a large, very lightly freckled egg into his hand and Glen withdraws it into the bag, where it is nice and dark. He gently places the candling torch against the end of the egg, illuminating the mesh of blood vessels inside. 'Fertile – about 25 days old,' he says, passing the egg back to Nichollette. The second egg is 23 days old. The eggs are carefully returned to the nest as Glen explains that both embryos

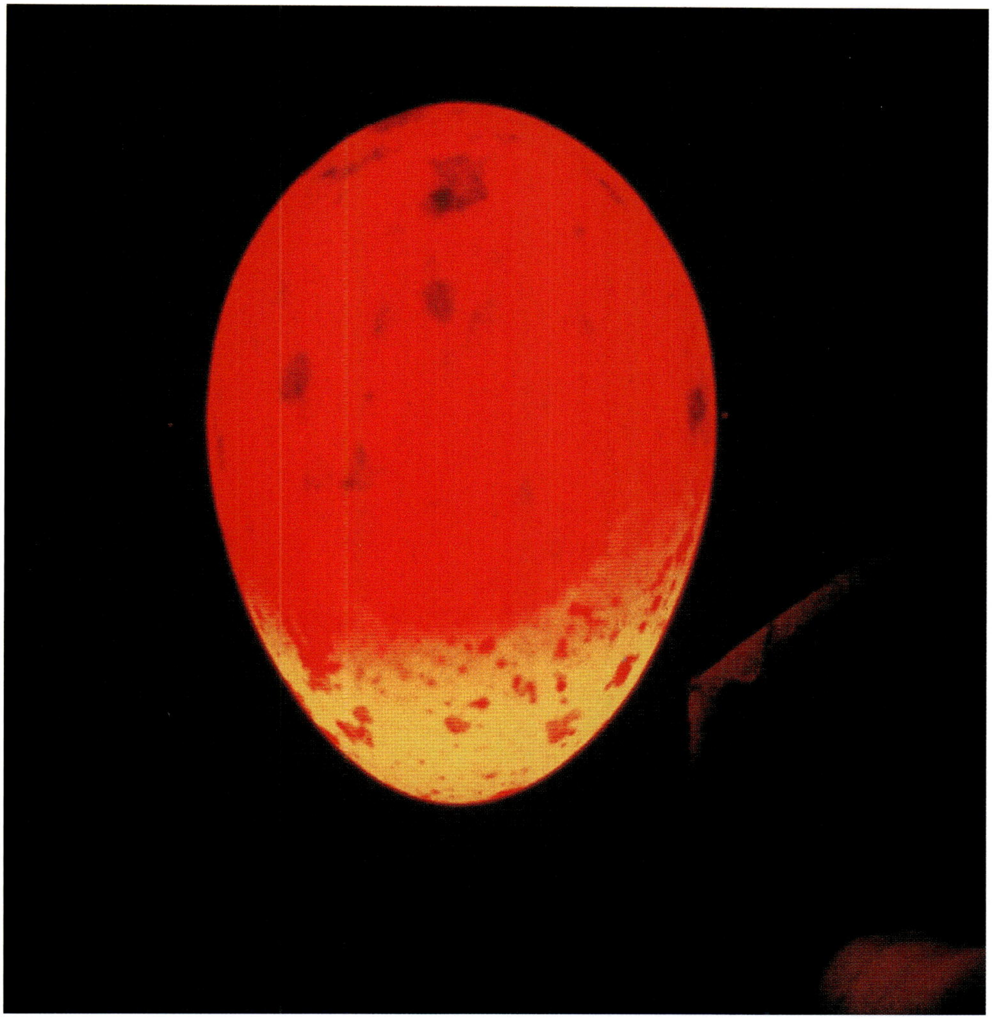

Candling involves shining a light through an egg to show its viability and development. This egg is fertile. DARYL EASON / DOC

were active and healthy and will hatch within a week. We quickly pack up and move away, to allow the incubating bird to return.

Glen and Nichollette are pleased the residents of this territory have two fertile eggs. Although it's unlikely that both chicks will survive and fledge, chances are good that one will make it through.

It's time to tackle the steep climb back up to the ridge and over into a small side valley leading down to Takahē Valley and back to the hut. We're not expecting to come across any more birds, but as we're dropping down the other side Yuki picks up a scent. We begin to get glimpses of a bird racing downhill, and Nichollette and Glen follow in hot pursuit. It's not easy negotiating the thick Dracophyllum *bushes and the uneven rocky ground, and I'm soon left way behind. The bird is making much better progress than its pursuers, and then it gets tangled up in a bush, giving Glen just enough time to catch up and grab it. By the time I reach them, Glen and Nichollette have identified the bird, which turns out to be wearing a radio transmitter. She is two-year-old Hakapoua, a parent-raised bird from the Burwood Takahē Centre. But she's a long way from where she was released the previous year, in the Dana Peaks area, and is clearly on the move in search of a mate. Hopefully she'll head down to where most of the Takahē Valley territories are, as she's more likely to find another singleton there.*

****

The takahē captive breeding programme was the first of its kind in New Zealand. It went on to become one of the longest-running and best-known endangered species captive breeding programmes in the world, but it began under a cloak of secrecy.

The early management approach to takahē was generally hands-off, leaving the birds to live and breed in their remote mountain hideout. However, there were growing concerns about the future of the small population, and some people were advocating bringing birds into captivity and trying to breed them.

Influential support for this idea came from British ornithologist Peter Scott, who visited New Zealand in the 1950s. Peter (incidentally, the son of Robert Falcon Scott of Antarctic fame) had established the Wildfowl and Wetlands Trust at Slimbridge in the United Kingdom in the 1940s. Since the early 1950s he had been involved in a last-ditch captive breeding programme for the nēnē or Hawaiian goose. The last 30 or so individuals of the species were brought into captivity in both Hawaiʻi and at Slimbridge, and over the next few years hundreds of captive-raised chicks were released back into the wild to create the basis of a population that now numbers in the thousands. This programme saved the nēnē from extinction. Peter advocated strongly for a captive breeding programme for takahē, and was also enthusiastic about the idea that the programme would mean that the New Zealand public would have the opportunity to see the birds.

While not everyone thought the programme was a good idea, it did have the support of enough people to set it in motion. In October 1956, the Rare Animals Advisory Committee endorsed a proposal from the Department of Internal Affairs Wildlife Branch to remove some eggs from Takahē Valley to hatch and rear.

The only person in New Zealand who had anything close to the requisite skills to do this was Elwyn Welch, an enthusiastic amateur ornithologist from the Wairarapa. Elwyn,

Elwyn Welch (right) designed special boxes to transport small takahē chicks and their bantam foster mothers from Fiordland to the Wairarapa. He was accompanied on the 1959 collecting trip by Gordon Williams. PETER MORRISON / DOC

who was in his early 30s, ran the family farm at Kelvin Grove on the northern flanks of Pukaha / Mt Bruce. He had always been mad keen on birds, and over the previous decade had developed skills as an aviculturalist, hand-rearing grey teal chicks.

Gordon Williams got in touch with Elwyn about the idea of raising takahē, and they hatched a complicated plan for how to do it. They were very secretive because they didn't want anyone to know about the project until it succeeded. DSIR scientists helped design some cannisters to hold the eggs while they were being moved, protecting them and maintaining them at a desirable temperature and humidity. In late November 1957, four takahē eggs were collected from Takahē Valley and sent to Elwyn in the Wairarapa, who had been training bantam hens to incubate pūkeko eggs and raise the chicks. The first clutch of two eggs failed to hatch. In the second clutch, one egg reached the pipping stage just before hatching, but then also died. The fourth chick hatched but survived only a couple of days.

As it wasn't clear what the incubation requirements for takahē eggs were, the new plan for spring 1958 was to collect takahē chicks instead. Elwyn decided to take bantam hens down to Fiordland so they could begin incubation straight away. To accustom the foster mothers to the journey ahead, he walked around his farm carrying the bantams in their travel boxes.

Everything was carried out in absolute secrecy. Elwyn, along with Gordon and photographer Peter Morrison from the National Film Unit, apparently travelled under assumed names and told no one what they were doing. They collected four chicks, each just a few days old – a photo taken by Peter outside the Takahē Valley Hut in December 1958 shows Gordon and Elwyn holding the small wooden travelling boxes. The chicks travelled back to the farm in these boxes, safely tucked under their foster bantams.

The two clutches each comprised a male and a female chick, and the siblings from the first clutch were named Prudence and Peter, while the second clutch became Patience and Paul. Patience survived for only five weeks, but with a lot of human help in addition to early bantam assistance, the other three chicks survived to independence. Pictures from the time show the takahē, as tame as chooks, wandering around Elwyn's feet on the lawn in front of his house, where they lived confined within a wire-mesh fence. There were no tussocks or shrubs to provide shade or shelter from passing hawks, so the youngsters were given a cloth tepee for shelter; with its bold striped interior the structure had a jaunty circus air about it.

Baby takahē are fed a range of insects for the first few weeks before they graduate to accepting suitably trimmed plants from their parents, so Elwyn created an ingenious device for catching blowflies. He also fed the chicks grain, which they happily ate from his hand.

In early 1959 the chicks and the captive-rearing programme were announced to the world, when the *Evening Post* newspaper was invited to take photos of the chicks. The images made their way around the world, including into the pages of the *Illustrated London News*, which continued to regularly feature takahē stories. A picture of the three chicks in the London weekly, sporting tiny plastic leg bands so that Elwyn could identify them, was accompanied by the caption 'Posing unafraid and full of blowflies six weeks after their capture from a remote New Zealand valley: the surviving takahe chicks'.[1] There were other photos, too: the blowfly trap, the tent and Gordon weighing one of the chicks in a cloth bag suspended from a small handheld balance. An interesting comment in the story says that 'the bantam foster mother continued to look after [the chicks] until one of the chicks, searching for food in its guardian's head-feathers, removed half her comb

Aviculturalist Elwyn Welch, with Rosemary and Gilston Carroll and one of the first hand-reared takahē. Photo taken in 1959 on the Welch farm, known as Kelvin Grove, at Mount Bruce near Masterton. WAIRARAPA ARCHIVE REF: 10-119/2-22

with its powerful beak'.

The New Zealand public loved the takahē story. When the birds at Kelvin Grove were put on display in 1960, more than 13,000 people visited over three weeks.

Prudence, Peter and Paul grew quickly – so much so, that at six months of age they were all about 50 per cent heavier than wild adult birds. They were judged to be overweight and had their daily rations reduced. A subsequent analysis showed they were also larger than wild birds, which accounted for their heavier weight, so being put on a diet probably wasn't helpful.

Spurred on by the success of rearing the three chicks, Gordon decided that it would be a good idea to add a female to the captive population that could be paired with the lone

Prudence, Peter and Paul were the first takahē chicks to be raised in captivity, in 1959.
EVENING POST COLLECTION / ALEXANDER TURNBULL LIBRARY REF: EP/1959/0172-F

male, Peter (his sister Prudence would be paired up with Paul). So in July 1959, Penny – an adult female who was known to have bred – was brought from Point Burn to the Wairarapa.

The problem was that the three hand-reared birds were very imprinted on Elwyn, and this severely impacted their ability to breed successfully. In their first breeding season together, in 1960/61, Prudence and Paul didn't produce any eggs, while hand-reared Peter failed to fertilise Penny's five eggs.

The following year, Gordon and Elwyn took a new approach 'to overcome the apathy between the males and the females'.[2] Instead of being kept in neighbouring pens, the two pairs were allowed to share a single large pen. The change had no effect on Prudence and Paul, who still produced no eggs. Peter and Penny produced four infertile eggs, one of which Penny laid in Prudence's nest. Rather surprisingly, there was no fighting between the pairs, who actually seemed to cooperate, taking turns to sit on each other's nests.

Just as things were gathering momentum with the captive breeding programme, Elwyn, a committed Christian, decided to move with his young family to Nigeria to undertake missionary work. By this time the Wildlife Branch was committed to the takahē captive breeding programme, and it was also keen to try rearing kākāpō, so it agreed to buy Kelvin Grove from Elwyn to be the base of a breeding programme for native birds. This was the beginning of the Mt Bruce Native Bird Reserve, which would become the Pukaha National Wildlife Centre. Elwyn left New Zealand in April 1961, but sadly he contracted a rare form of polio and died in December that year.

According to Brian Reid, who started working as a scientist in the Takahē Recovery Programme several years later, the responsibility for looking after the captive takahē after Elwyn's departure fell first to photographer and ex-deer-culler Peter Morrison, who was made temporary caretaker. When he pulled out, the Wildlife Branch advertised for a caretaker and appointed Roy Walker, whose background was as a plumber. Brian says that although Roy liked birds and was diligent, he was rather disinterested.

Looking back, Brian admits he was rather horrified that the captive breeding programme for these 'absolute gems of New Zealand avifauna' got the initial go-ahead.[3] 'Who knew anything about birds in the Wildlife Branch?' he said. 'We were all deer-cullers.' In his opinion, 'these birds were too valuable to go to Fort Knox, yet we gave them to Elwyn'.

We'll come to Brian and the ongoing efforts at Pukaha / Mt Bruce to breed takahē in captivity, but first, let's go back to that comment about deer-cullers.

A bantam foster mother keeps a close eye on her young takahē charges at Elwyn Welch's farm, 1959. EVENING POST COLLECTION / ALEXANDER TURNBULL LIBRARY REF: PACOLL-10563-179

CHAPTER EIGHT

DEER YEARS

DEER IN NEW ZEALAND – THE BIG PICTURE

Early European settlers in Aotearoa rarely valued its native forest and wildlife, and yearned instead to create an agricultural paradise that reminded them of their homeland. To achieve that they brought with them a multitude of species from the northern hemisphere. Some, like sparrows, they brought to remind them of home. Others, such as trout and deer, were to stock the rivers and land with food and fur, as well as to provide sport and trophies.

The first red deer to be introduced to New Zealand arrived from England in the mid-1850s. They were soon liberated in forests across New Zealand, where they were successful well beyond the dreams of the newly created local acclimatisation societies. In England, hunting was the preserve of the well-to-do gentry, but in the new egalitarian society that was springing up in New Zealand, it was to be available to anyone.

Other species followed. Among them was the wapiti or North American elk, released into Te Houhou / George Sound in Fiordland in 1905. Some of these animals were a gift from United States President Theodore Roosevelt to the people of New Zealand.

At first, the deer were protected and managed for hunting by the acclimatisation societies, which were overseen by the Department of Internal Affairs. It wasn't long, however, before rapidly growing red deer herds were having a significant impact on forests, stripping the undergrowth and preventing regeneration. Erosion and flooding soon followed. By the early 1920s, it was estimated there were more than 300,000 naturalised deer in New Zealand, and that number was growing at an alarming rate. Herds of starving deer began to invade farmland in search of food.

Under the Animals Protection and Game Act 1921, responsibility for managing deer numbers fell to Internal Affairs, and in 1930 the department held the Deer Menace Conference in Christchurch, which was also attended by Forest Service staff and conservationists.

Red deer are widespread across New Zealand. **MATT WINTER**

Protection for deer was lifted and a price was put on their head, and later that year government hunters employed by Internal Affairs began a widespread culling programme. The aim of the deer destruction campaign was extermination.

More than 11,000 deer were shot in 1932, and that number climbed year on year. By 1940, when more than a hundred men were working as cullers, the total stood at 30,000 deer a year. By 1944, with the New Zealand Army helping out to replace hunters who were away fighting in the Second World War, the overall tally of deer killed reached 250,000. But despite this effort, it was clear that deer numbers were still rising rapidly.

In 1945, the new Wildlife Branch of the Department of Internal Affairs was established. This was responsible for deer control, along with fish acclimatisation work in the Rotorua and Southern Lakes districts and also the protection of some native birds. For the first time, scientists were brought in to investigate a range of issues, especially the impact of the rapidly growing possum problem on native forests.

The deer problem had become a topical issue, both politically and with the New Zealand public, who were developing an appreciation for native plants and animals. In January 1949,

Large numbers of deer in the Murchison Mountains in the 1950s and 1960s caused significant damage to the alpine grassland and the understorey of the beech forest. WILLIAM LEE

the Wildlife Branch found itself supporting what was at the time the largest scientific expedition in New Zealand. The New Zealand–American Fiordland Expedition involved more than 60 people based between Taitetimu / Caswell Sound and Te Houhou / George Sound, whose aim was to study the wapiti and how they were adapting to their new environment. This added impetus to a growing move away from the concept of deer extermination to deer control, and supported the idea that a moderate wapiti population could – and should – be maintained in Fiordland for trophy hunters to target.

Surprisingly, there had been little research to quantify the scale of the deer problem at a national level, and it was only in 1951 that a newly appointed ecologist in the Wildlife Branch, Thane Riney, began to collect evidence that the deer destruction campaign had been far less effective than claimed. Culling efforts had focused on areas with high deer numbers, and this produced correspondingly high kill rates. Thane suggested instead that effort should be focused on critical areas where the forest was being particularly hard hit or there was bad erosion. This would be a much-debated suggestion until 1956, when deer-culling and other wild animal control operations were moved from Internal Affairs to the Forest Service.

DEER IN THE MURCHISON MOUNTAINS

The discovery of takahē in November 1948 came against this background of growing deer numbers and a large team of cullers working on foot around the country. It also coincided with evidence that red deer were spreading rapidly in the Murchison Mountains.

Red deer had been released near Manapouri, about 30 kilometres south of the Takahē Special Area, between 1901 and 1910. The Murchisons are effectively a large peninsula, surrounded on three sides by Lake Te Anau, and as a result deer invaded the range much later than many other parts of Fiordland. According to experts such as mammal ecologist John Parkes, the deer reached the Murchies 'via the Kepler Range and around the head of the South Fiord of Lake Te Anau, arriving in western catchments … in the late 1920s'.[1] By the early 1940s, wapiti had also arrived from the north, but they failed to thrive in the area.

Red deer populations follow a predictable trend when they arrive in a new area: following their initial establishment, their numbers explode, hitting peak density after about 20 years. In the case of the Murchisons, 'at the time of the rediscovery of the takahē in 1948, deer were at peak numbers in parts of the northwest Murchison Mountains, and rapidly beginning to colonise some of the eastern and southern catchments'.[2] The situation was taken seriously at the highest level.

A tussock that has been heavily browsed by deer will take decades to fully recover.
WILLIAM BROCKELSBY / DOC

On 23 April 1951, Secretary for Internal Affairs Arthur Harper wrote to the Ornithological Society of New Zealand outlining the measures that were being taken to protect takahē (the letter was published in the special 1951 edition of *Notornis*, the newly renamed journal of the society). The letter made it clear that, in addition to establishing the Takahē Special Area and protecting the birds, one of the first priorities was to control the numbers of stoats, deer and possums:

> [Frank] Woodrow has carried out control work with respect to the introduced animals found in the area. This has involved the destruction of stoats by trapping in such manner that there is no likelihood of the Notornis or other ground birds being caught; the snaring of deer so that there is no possibility of the birds being disturbed by shooting and finally the trapping of opossums [*sic*] in the restricted areas where they occur. Fortunately, there is no evidence that either rats or cats exist in the known Notornis range.[3]

A year later, Gordon Williams, also writing in *Notornis*, had the following to say about deer:

> Apart from the danger that may exist during the birds' breeding season of deer treading on nest or chicks, there is perhaps a greater danger of increased or even uncontrolled deer herds so altering the environment in the subalpine meadows where the birds feed for most of the year and make their nests, that food and cover will eventually be greatly diminished. Already in the northern basin that forms part of the great amphitheatre at the head of the Point Burn there is evidence that the presence of deer is accelerating the normal processes of erosion … Lower down on the boggy glacial benches, there are places where the thick covering of the various grasses and herbs has been totally destroyed over many square yards and turned into muddy deer wallows. Deer feed on species … that the takahe have been shown to depend upon in large part for food and nesting cover.[4]

While there were still only small numbers of red deer in Takahē Valley and Point Burn in 1948, this changed significantly over the next decade, with mobs of deer becoming a common sight throughout the high basins of the Murchisons.

From 1953 onwards, Frank Woodrow's deer control work in the Murchisons was supplemented by the efforts of official government deer-cullers. That year, the local Internal Affairs field officer reported that cullers were each killing about 30 deer per day in the northwest, in valleys such as the Snag Burn, with a total of 350 deer shot. The following year the total was 517. From 1955 to 1962, cullers killed 4,363 red deer and 88 wapiti in the Murchison Mountains, and it's estimated that recreational shooters killed a further 1,500 or so. Deer numbers, however, remained high.

In 1962, the Forest Service took over deer control in the Murchisons. The following year saw a whopping 1,767 deer killed, at an average of nearly seven per hunter. By the end of the decade the total number of deer killed by the Forest Service shooters was more than 8,000.

The impact of the culling was to prevent deer numbers in the eastern valleys of the Murchisons peaking as they had done naturally further west, but it was clear that the deer were still having an impact on the vegetation. The question was, how much of an impact was this having on the takahē?

Home away from home

Up to 26 huts have existed in the Murchison Mountains at one time or another, providing the rudiments of comfort and shelter from Fiordland's extreme weather. Unlike most other huts on the New Zealand conservation estate, these structures lie in an area closed to the public – the Takahē Special Area. They are for the exclusive use of DOC staff, researchers, deer-cullers and predator trappers. They are mostly small and very basic – a couple of bunks and a cooker, usually in a space measuring no more than 10 square metres. Some huts have an unlined annexe for storing raincoats, boots and packs out of the rain and away from prying kea eyes, and a few in the forest have – or had – fireplaces.

Takahē Valley Hut, built in 1954, is the oldest surviving hut, although Doc Orbell's log cabin, which he built in the Ettrick Burn sometime in the early 1950s, may have been the very first hut. The third iteration of the 'log cabin' (with no logs in sight) was built in 1999 and is the newest hut in the Murchies.

Wildlife Service trainees helped build a basic slab hut in the Point Burn Valley in 1959, but heritage consultant Michael Kelly, who wrote a comprehensive report on the history and condition of the huts in 2010, says building took off in 1961, when the Fiordland National Park Board and the Forest Service began to construct a network of huts across the Murchison Mountains, mainly for deer-cullers to use. First was two-bunk Wisely Hut, named for Canterbury University zoologist Baughan Wisely. Early in 1962, Geoffrey Orbell and others were commissioned by the park board to build a hut at Island Creek, near Miller Peak. In 1963, the Forest Service began to

erect their standard-design prefabricated huts, sometimes at the rate of four a year. Getting building materials into the Murchisons was a significant undertaking. Michael writes that 'Aurora Point Hut was built close to [Lake Te Anau], so the opportunity was taken to ship the prefabricated sections in by boat – the *Tawera* – prior to its construction.'[5] Materials for huts at Lake Eyles and Lake Te Au were flown in by floatplane. 'The remainder of the huts were flown in by Dominies [a transport plane used during the Second World War], with the materials tied in bundles and parachuted to the ground.' In 1965, the legendary pilot Bill 'Blackie' Black was the first to fly loads of prefabricated hut material by helicopter.

Hut building paused in 1967, and then began again in the mid-1970s with helicopters flying complete lightweight, relocatable aluminium bivvies and dropping them into position. This second stage of hut building lasted until the mid-1980s and was carried out by both the Wildlife Service and Forest Service.

Before an adequate number of huts was in place by the late 1960s, deer-cullers and researchers set up tent camps and bivouacs, and even used rock shelters. These days, most deer-culling is done by helicopter, but people servicing the stoat traps in the Murchison Mountains continue to rely on those small historic huts for sanctuary from the elements. Michael writes: 'The huts themselves are, like tracks and bridges, a physical manifestation of government effort to protect the biodiversity of these places. Together they tell the story of human effort expended to save threatened species and exceptional habitats.'[6]

OPPOSITE ABOVE: Wildlife Service officer Bill Axbey in 1971 standing outside the slab timber hut built in 1959 in the Point Burn valley, Murchison Mountains. The building iron was air-dropped from a Dominie biplane. **R.H. SIMPSON / DOC**

OPPOSITE BELOW: Small (2 metres by 2.5 metres) two-bed bivvies that can be relocated by helicopter are a staple field accomodation in the Murchison Mountains. Former takahē ranger Nick Torr says huts like the one in Waterfall Creek are "well insulated and cosy, and a thousand times better than a tent." **NICK TORR**

CHAPTER NINE

EARLY DAYS AT PUKAHA / MT BRUCE

In the early 1960s the deer-cullers were being kept busy in the Murchisons, but it was a quiet time there for takahē research. Following an intense decade studying the basic biology of the takahē and trying to work out the size of the population in the Murchisons, the hope now was that captive breeding would be a way of growing the population, and that managing deer would be enough to allow the wild birds to thrive.

When Brian Reid joined its small research section in April 1963, the Wildlife Service had few people or resources available. Gordon Williams, who had been the driving force behind research in the 1950s, had left the Wildlife Service and moved to Lincoln College (he would return several years later). Brian recalls that much of the equipment being used at Pukaha / Mt Bruce was second-hand, and that there was a lack of funds and support.

There were still four takahē at Pukaha / Mt Bruce: Paul, Penny, and siblings Prudence and Peter. Following the first two unsuccessful breeding seasons, the pairs had been swapped around, resulting in brother and sister Prudence and Peter being paired together. This genetically undesirable pairing failed to produce any eggs, while Paul and Penny produced five infertile eggs.

The Wildlife Service decided to add another breeding pair to the captive population. In September 1963, Brian travelled down to Takahē Valley and returned with an adult pair of takahē, named Penelope and Patrick. These early translocations were hard on the birds – this was before the days of helicopters, and they were usually backpacked out of the mountains, before travelling by boat, plane and finally car to reach the Wairarapa.

Within a month, Penelope had died from septicaemia that began with a toe damaged in the capture and relocation process. It was too late to pair Patrick with either of the Pukaha /

Colin Roderick, in 1977, with Shirley, Jock and their male chick (centre). The unnamed chick remained at Pukaha / Mt Bruce, where he fathered five offspring with his partner Red-Left. These became founder birds on Mana Island. ARCHIVES NZ - COMMUNICATE NZ COLLECTION R24814857

Mt Bruce females, as they were already breeding – unfortunately, with exactly the same poor results as the previous year.

The years ticked by with no success. In the 1964/65 breeding season, Patrick was paired with Penny, but once again she laid five infertile eggs. Prudence, Peter and Paul were left together in what Brian described as an 'eternal triangle' in the hope that might boost success. However, the two males ignored Prudence and instead spent a great deal of time together, feeding and preening. Peter was treated with gonadotrophic hormone in an effort to bring him into breeding condition; it had no effect. Prudence didn't lay any eggs.

It's no surprise that birds from the Murchison Mountains struggled in the Wairarapa. Instead of a cool, wet climate and plenty of tussock, they were faced with a paddock of pasture grass and a climate that in summer was very hot and dry. In 1965, Brian and recent wildlife trainee Colin Roderick, who had taken over the breeding programme when Roy Walker was dismissed after falling out with his colleagues, took the advice of a visiting avian expert to improve the bare grassy enclosure by planting it with tussocks. Brian and Colin travelled to Waiouru, in the central North Island, where they dug out the first 1,500 of what would end up being 5,000 tussocks. Back at Pukaha / Mt Bruce, they were faced with the equally monumental task of planting them all out.

In July 1965, Peter died and Prudence was reunited with Paul. The result was predictable – no eggs. Then, early in the 1965/66 breeding season, Penny also died, leaving Patrick a bachelor. An autopsy was carried out and Penny's cause of death was attributed to old age and visceral gout, a disease unique to birds in which kidney failure causes a build-up of uric acid crystals in the internal organs.

With the captive population down to one female and two males, the Wildlife Service decided to add another wild pair to the programme. The next two birds were Thisbe and her partner Pyramus, collected in October 1966 from the Ettrick Faces. In *Metamorphoses* by the Roman poet Ovid, Thisbe and Pyramus were a pair of ill-fated lovers whose story was an early version of Romeo and Juliet. Their takahē namesakes didn't fare much better. After less than three months, Pyramus died from what was described as a 'chill'. He hadn't adapted well to Pukaha / Mt Bruce and his weight had dropped significantly since he was caught. An unseasonal cold snap with continuous sleet and hailstorms proved too much for Pyramus, and he died from congested lungs.

Despite the ongoing setbacks, the Wildlife Service was determined to make captive breeding work. Just four months later, in February 1967, it was back in Takahē Valley. This time, the team caught a breeding pair and their two chicks, aged just a few weeks. This did not go well – on their first night at Pukaha / Mt Bruce, both chicks were killed by their parents, who were extremely stressed by the move. A week later, the female, Monash, died, although her mate Jock would live for a further 17 years. Later that year, Paul also died, having survived six-and-a-half years since he was moved as a chick. A couple of months later, Thisbe died, following an ill-conceived attempt to artificially inseminate her. This broke an egg in her oviduct, resulting in a fatal infection. Patrick died in late 1968. The only surviving takahē were Prudence and Jock, who were relocated to Kapiti Island for four years from mid-1968, leaving Pukaha / Mt Bruce free of takahē for the first time in a decade – but not for long.

In 1970, the Wildlife Service decided to rejuvenate the moribund captive breeding programme at Pukaha / Mt Bruce, and also attempt a further translocation of wild birds to Kapiti Island to see if that would work as a home for takahē. In January, a pair was caught at Lake Eyles and moved to Kapiti via Pukaha / Mt Bruce, but both birds died within a fortnight. The cause of death was starvation – the stressed birds simply had not been eating enough.

In February 1970, four more birds were moved from Fiordland to Pukaha / Mt Bruce. Winnie and Pooh were a breeding pair, while Modok and Gretchen were father and daughter. Winnie came to an unfortunate end six months later when she tunnelled underneath a fence from her paddock into Gretchen's territory. During a fight with the fiercely territorial

This 1984 photo of Jock may be a little blurry, but it is an important record of the first bird recorded on the takahē pedigree. Jock and partner Shirley were moved to the Te Anau Wildlife Park in the early 1980s after many years spent at Pukaha / Mt Bruce. ALAN WRIGHT / DOC

Gretchen, she received fatal head injuries. Gretchen and Modok were left together to breed, but – not surprisingly, given their familial relationship – failed to establish a pair bond and for two years in a row Gretchen laid infertile eggs.

By mid-1972 the captive breeding programme had been running for more than 13 years and had failed to produce a single chick. Of the 17 birds moved to Pukaha / Mt Bruce during that period, 12 had died.

But the Wildlife Service was still not prepared to admit defeat. This was a time of renewed effort to work on takahē, spurred on by news of a declining population in Fiordland. And Brian had come up with a way that he hoped would alleviate the stress that pairs of takahē were experiencing through living in close proximity with one another without being able to resolve their territorial differences. It was this that he and Colin believed was preventing successful breeding. The two adjacent pens were now just quarter of a hectare each, as the bush areas had been fenced off to create pens for captive kiwi, bringing the takahē pairs into close proximity. In contrast, wild birds have much larger home ranges in the Murchison Mountains and can effectively avoid other birds during the breeding season by limiting themselves to a small territory. Outside the breeding season they range further afield and even overlap amicably with neighbouring pairs at this time.

The birds at Pukaha / Mt Bruce were able to hear one another all the time, although they were separated by a fence that had changed over time from a wire-mesh to a solid wooden structure so they couldn't see one another. The fence seemed to be part of the problem, as it prevented warring birds from fighting to resolve territorial ambitions, leaving each male believing his territory included both pens. Putting gates in the fence and allowing the birds access to the other pair simply resulted in long, intense fights. In addition, the loser had nowhere to retreat to, as the winner kept chasing him, believing he was still trespassing.

In an attempt to overcome these problems, Brian invented what he called the 'neutral' zone, which was a small intermediary zone both pairs could access but that gave them only a limited view into the adjacent pen. Sparring males would still fight, but the winner was now less confident about chasing the loser. Over time, the birds accepted the pen they were housed in as their territory and lost the inclination to enter the other pen.

The modified set-up was put to the test in 1972. Prudence and Jock were brought back from Kapiti Island and placed in one pen, and Gretchen was paired up with Pooh, while her father, Modok, was surplus to requirement and housed elsewhere. Effectively, the birds could move between pens as they wished. Prudence laid a miniature soft egg without a shell, while Gretchen laid two infertile eggs. But when the eggs were removed, Gretchen and Pooh bred again – she laid another two eggs, one of which was fertile but died early in development. Their third clutch was a milestone – two fertile eggs, both of which hatched. Unfortunately, both chicks had development issues and died within a few days, but there was renewed hope that captive breeding might work after all.

On the basis of this, in April 1973 another wild pair was collected from the Murchison Mountains – Elwyn and Shirley, named after the Welches. But Elwyn was permanently crippled after being treated with drugs soon after his arrival at Pukaha / Mt Bruce, which made him unable to mate, and the next breeding season had barely begun before Pooh and Modok died after fighting with Jock, despite the presence of the neutral zone. They both succumbed to a blood condition caused by a *Campylobacter* bacterial infection. Following that incident the team decided that a solid gate between the two pens, rather than the neutral zone, would be advisable. These deaths and fears of an infectious disease would have strong implications for the movement of takahē between different sites for the next few years.

Over the next two seasons, the breeding programme stuttered on. Shirley and Elwyn laid infertile eggs, while Gretchen and Jock managed one fertile egg each season, neither of which hatched. Gretchen died in late 1974. In April 1975, a further wild pair, known simply as the Plateau Pair, were added to the mix, bringing the total number of birds collected from Fiordland and moved to Pukaha / Mt Bruce to 23. It was a lot of birds with not much to show for it.

Pukaha / Mt Bruce in the 1980s

There was a second round of successful captive breeding at Pukaha / Mt Bruce, in the mid-1980s. The unnamed male chick born to Shirley and Jock in 1977 went on to father five chicks with a female identified in the takahē pedigree as Red-Left, a wild-born bird who may have been the female of the Plateau Pair. These two raised a male chick called Taku in 1984, the female Georgie and male White in 1985, and two further females, Mataku in 1986 and Betty in 1987. These birds would go on to become island founders, and Taku in particular would play a pivotal role.

However, this small flurry of breeding was a case of too little, too late. In 1988 it was decided that Pukaha / Mt Bruce would no longer be considered as a breeding site for takahē. The centre would continue to hold one pair of birds on public display, as the takahē had been a popular exhibit since the late 1950s, being the only place in the North Island where these famous – and rare – birds could be seen.

In the 1975/76 breeding season there was another glimmer of success. Shirley and Jock laid two fertile eggs and the single chick that hatched made it to 16 days before dying.

The breeding programme's first real success finally came in late 1976. Shirley and Jock produced four fertile eggs, and two chicks hatched and – for the first time – survived. The two females were called – in the fine 'P' tradition of Pukaha / Mt Bruce takahē – Precious and Perfect. They would survive until five and six years old, respectively, although one was partially crippled by a leg and toe abnormality.

In 1977 Shirley and Jock produced two more chicks, one of which survived. This unnamed male would live to be 12 years old. And in 1978 they produced another survivor, a male named Junior.

This period proved to be some of Pukaha / Mt Bruce's most successful years, although neither Precious, Perfect nor Junior would go on to breed. However, there would soon be a significant change in direction for the captive breeding programme that would see the emphasis shift from the Wairarapa back to Southland.

CHAPTER TEN

HOW MANY TAKAHĒ?

Day 2 of the 2017 Murchison Mountains head count finds me with Glen Greaves, Nichollette Brown and dog Yuki searching up Takahē Valley, recording how many birds we can locate. In mid-afternoon, at the head of the valley at more than 1,500 metres altitude, where there is still some snow on the ground, we come across a family group: mum, dad, and a subadult born the previous year and sticking around to help its parents during the imminent breeding season. The birds move away from us, but they don't break into a run. The vegetation up here is sparse and low, so it's easy to see their legs. Through their binoculars, Glen and Nichollette can see that one bird has leg bands: a red plastic band sitting above the numbered metal band (recorded as 'RM') on its left leg, and a red plastic band above a yellow plastic band (RY) on the right leg. A quick check reveals that RM-RY is a four-year-old male named Parker, a genetically valuable male because of his wild heritage. Parker was born to wild parents in the Murchisons and he has paired with another wild but unbanded female.

Glen decides that it would be useful to band the subadult and fit it with a radio transmitter, to add another wild takahē to the radio-tagged and regularly monitored sample of birds. It is quite an easy catch, and while Glen sits with the bird in his lap, head and strong legs carefully restrained, Nichollette sets about attaching bands to its legs and fitting a radio transmitter to its back. They decide to name the 2.4-kilogram youngster (later revealed to be female) Telfer, which is the maiden name of Joan Watson, a member of the 1948 rediscovery party. While the pair work, the parents hang around nearby, making anxious booming contact calls, and when Telfer is released, the reunited family scuttles away. The adults clearly aren't breeding yet, which Glen attributes to the late snow covering the ground.

It was a very successful takahē head count. Elsewhere in the Murchies, another subadult was fitted with a radio transmitter and named Luminosa, after the scientific name for the glow-worms that are famously found in the nearby Te Ana-au Caves. Burwood-raised

Nichollette Brown fits a metal leg band to a young bird held by Glen Greaves. The takahē was named Telfer after Joan Telfer, who was part of the 1948 rediscovery team. Telfer was caught at the head of Takahē Valley during the 2017 Murchison Mountains census. Radio transmitters allow a subset of birds to be tracked from a plane as part of ongoing population monitoring. **ALISON BALLANCE**

Korokoro, who had lost her transmitter, was given a new one. And excitingly, Tintooki, who hadn't been seen for 10 years, was spotted – a reminder that a few birds would have escaped detection despite the intense search effort.

After 70 person-days in the mountains, the searchers between them counted 118 birds. As a few of the more remote birds were bound to have been missed, the takahē team were satisfied that the ground-truthing provided strong support for the statistical estimate of the population from the radio-tracking study, which was 130. We saw 17 birds in Takahē Valley and the nearby valley we first searched, which was pleasingly up from 13 counted three years earlier. Overall, lots of takahē were in the early stages of nest-building, while there were 10 nests with eggs – and an early-bird family that had already hatched a small chick. The takahē team were very happy – as well as rather tired and sunburnt from the unseasonably nice weather.

****

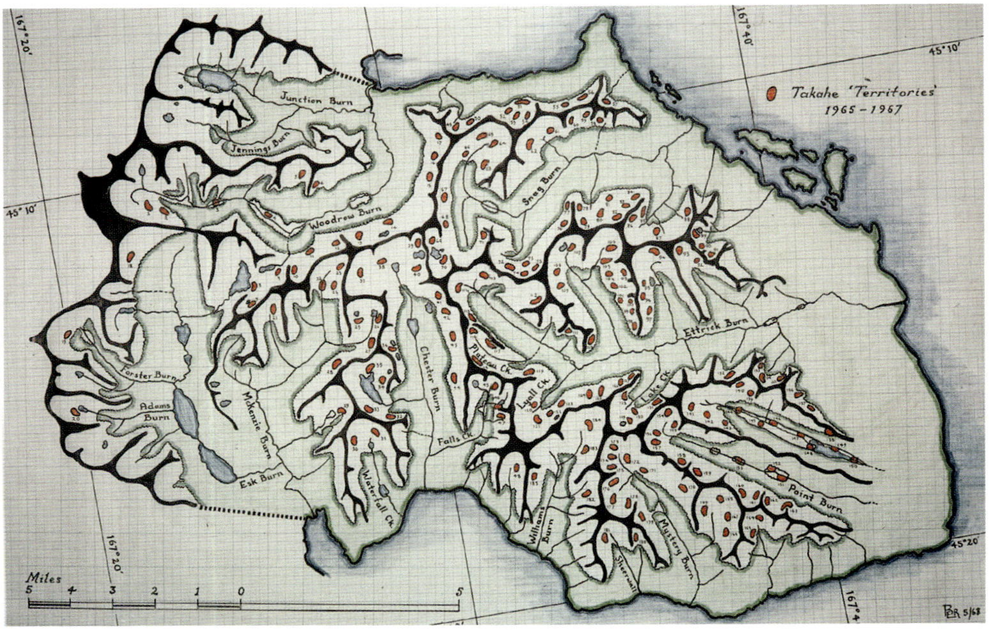

This map of takahē territories in the Murchison Mountains was compiled between 1965 and 1967 and relied heavily on sightings from observant deer cullers as well as Wildlife Service staff. DOC

From the moment takahē were rediscovered in 1948, the question on everyone's mind was, how big is the population? It would take just over 30 years to get the first accurate estimate, by which time there were just over a hundred birds remaining in the Murchisons, with a handful on some outlying islands. It took that long because of the remoteness and ruggedness of their last stronghold, a lack of resources, the small numbers of searchers and the usual inclement Fiordland weather. The situation was compounded by a confusion about the profusion of feeding sign that the birds leave in their wake as they roam over sometimes large distances, creating the illusion of many more birds than there actually are.

Takahē Valley was the site of the rediscovery, and it and the nearby Point Burn Valley remained the focus of surveys and research for several decades. The valleys were relatively easy to access from Te Anau, and their topography was benign compared to that of many other catchments, especially those further west. The baseline population monitoring set-up in these two valleys would later prove pivotal to the realisation that takahē numbers were in freefall.

We don't know how big the population was at the time of the rediscovery, but there was soon evidence that it was shrinking, with birds in the westernmost valleys disappearing first. In 1960, after the first decade of intensive research, Gordon Williams estimated there were a few hundred takahē in total, with a steady 10 pairs in Takahē Valley and Point Burn.

It was thought there were 12 territories in total in those two valleys, but several of these were usually empty. Including juveniles and transients in this number brought the estimated population there to 28–30 birds. This number was only for the lower reaches of the two valleys and didn't include the steep head basins, which were later found to be home to at least eight pairs, bringing the total population in the study area to about 47–50 birds.

For several breeding seasons from 1960/61 on, researchers made few visits to the Takahē Valley and Point Burn study area. The Wildlife Branch staff had decided it was better to minimise disturbance to breeding birds, so the banding programme for adult birds was in abeyance, although a few chicks were banded.

Banding and more intense surveying recommenced in the study area in the 1963/64 breeding season. This was prompted by Brian Reid's concern that, based on previous observations of banded birds, the mismatch between high mortality of adult birds and low survival and recruitment of chicks suggested the population wasn't even replacing itself, let alone growing. Banding was continued in the 1964/65 and 1966/67 breeding seasons, and by the end a total of 71 adults and 33 chicks had been banded. Brian was now satisfied that the breeding population in Takahē Valley and Point Burn was stable, despite the removal of eight adults to Pukaha / Mt Bruce. After conducting a broad population survey of the western Murchisons in 1966 and the eastern catchments the following year, he estimated the total population stood at 370–420 birds. It later became apparent that this number was overly optimistic, and the takahē population was more likely to include between 213 and 243 birds.

But Brian was concerned that the population had been dwindling for many decades, and he urged people not to be complacent. In 1967, he wrote:

> there is no basis for expecting a declining and small population to stabilise at its present level particularly as this flightless vegetarian's 'natural habitat' is now shared with browsing ruminants and predatory stoats. (During the 1966/67 season one chick and one adult were killed by stoats, one nest was predated and another adult was probably killed by a stoat in Takahē Valley alone.)[1]

At the same time as the research focus on Takahē Valley and Point Burn was stepped up, staff from the deer control programme were undertaking regular expeditions to the Murchisons and making a big effort to note where they were seeing birds, feeding sign and possible territories. Two Forest Service cullers made a particularly notable contribution. From late 1964 until early 1968, Richard (Dick) Deaker made 153 reports, including 48 sightings of new pairs that hadn't been recorded before. His colleague Russell Dawson made a further 67 observations in the late 1960s. There was, however, a caveat to their reports. Despite being excellent observers, the shooters visited only some areas – since their focus was deer control, they preferentially went to places where they were likely to encounter deer, leaving some areas uncounted.

Dick made another vital contribution to the takahē programme. In 1964, he began marking territories and each year's sightings of birds on a map. This greatly clarified where birds were being seen, and this annual plotting became a routine part of data collection in the Takahē Recovery Programme.

But it was the annual late-February counts in the Takahē Valley–Point Burn study area that caught the attention of the Wildlife Service. From the steady 47 or so birds that were reliably counted in the early 1960s, numbers started dropping quite precipitously. Brian reported 36 birds in the 1967/68 season, 26 in 1968/69 and just 22 in 1969/70. Most of the birds had disappeared from the flat valley floors. Gordon Williams, who had returned to the Wildlife Service in 1965 as the head of research (he would later become its director) was initially not convinced by the numbers and sent down a team to carry out a survey. With numbers in the 1970/71 breeding season dropping to 18, alarm bells were really ringing. It was clear that if this trend continued, takahē would soon be extinct.

No one was sure why numbers were dropping so fast. Various ideas were put forward: predation by stoats; competition with deer, which had damaged or destroyed large areas of favoured takahē food; changing climate; and even more way-out suggestions, such as anthrax. These concerns and uncertainties would usher in a new era of takahē research and management.

ABOVE: Young wild takahē. **ANDREW DIGBY**

OPPOSITE: A wild takahē in the head basin of the Ettrick Burn. **ROSS CURTIS**

Takahē – place in the bird world

Takahē are classified as rails. Rails are ground-living birds that share a basic body plan of a solid, compact body; strong legs; short, rounded wings; and a short, often cocked tail. Rails belong to an order of birds known as Gruiformes – gruiform means 'crane-like', and the group is a bit of a grab-bag of cranes, rails and a few other species in search of a taxonomic home. The Gruiformes belong to the superfamily Ralloidea, and are divided into coots, crakes and gallinules, with takahē and pūkeko classed in the latter group.

Takahē versus pūkeko

Although pūkeko and takahē are classified in the same genus, *Porphyrio*, the fact that one species can fly and the other can't is just the beginning of the differences between the two. Physically, takahē are taller, heavier and more solid, with a big head and massive bill, and stout legs. Pūkeko are altogether smaller and more finely featured. Takahē are dark blue with a greenish back, and pūkeko are violet-blue with a black back.

Socially, takahē can be described as sedate, even staid, birds. They are dutiful parents with, usually, just one baby that they dote over. And even when that baby is a young adult, it often refuses to leave home, preferring instead to stay with mum and dad for a year and help out by babysitting. Takahē are serially monogamous suburbanites, even if their suburb of choice is a remote mountain range – or is that simply a rural lifestyle block with an exceptional view?

Pūkeko, on the other hand, are flamboyant and polygamous, with large, rowdy families of siblings and half-siblings, all raised in a hippy-like shared, open-parenting arrangement. The adults are quarrelsome squabblers, always scrapping with the neighbours. They are fond of a good wetland, but equally at home in a big city hanging out on the side of a busy motorway. Basically, they are happy to hustle for a living and they're not fussy.

The genus *Porphyrio* is very widespread geographically, with members on all continents except Antarctica, as well as on many islands, including across the Pacific. It is a group of 16 species (10 living and six extinct) that have been variously regarded at different times as either full species or subspecies. The North Island takahē and South Island takahē are two species, while the pūkeko belongs to a 'complex' known collectively as purple swamphens, which is currently considered to contain six species.

Swamphens are seen as one of the 'supertramps' of the bird world. The American biologist

A pukeko and a takahē graze side-by-side on Maud Island. The pair's evolutionary relatedness is very clear, but the takahē is clearly a much larger, more robust bird with quite different feather colours. **DAVID BOYLE**

Jared Diamond coined this term in the early 1970s, after the rock band Supertramp. An avian supertramp is a species that moves often and easily. Swamphens are both opportunists and generalists, able to take advantage of a wide range of food and habitats. They can also fly across long stretches of sea. Like other swamphens, the pūkeko is also a supertramp, unlikely though that may seem – once it manages to lurch into the air, it's an ungainly flier, with neck and head stretched out and legs dangling.

The pūkeko is thought to have arrived and established in New Zealand just 500 or so years ago. By that time, early Polynesian settlers had already begun changing some of the New Zealand landscape through widespread burning, creating habitats that were more conducive to a bird of open wetlands than the previous forest would have been.

Pūkeko are still in the habit of turning up occasionally on isolated islands around New Zealand. One was recorded on the Snares Islands / Tini Heke in 2016, while back in 1947 another made it all the way to subantarctic Campbell Island / Motu Ihupuku, about 650 kilometres south of Invercargill. There is also a small population on Raoul Island in Rangitāhua / Kermadec Islands.

Ancestry

Exactly how closely related are takahē and pūkeko? Although they shared a common ancestor a few million years ago, they are not sister species – more like first cousins once removed, perhaps. To be sister species they would have to be descended from the same ancestral population, which was what scientists initially assumed. Under this scenario, individuals of the same species dispersed across the Tasman to New Zealand on three different occasions. By the time of each subsequent arrival, the previous invasion had evolved into a new niche and was no longer capable of interbreeding. The first arrival gave us the South Island takahē and the second resulted in the North Island takahē, while the pūkeko is the most recent arrival.

However, more recent genetic evidence from the Phoenix Lab at Massey University suggests that the ancestral birds weren't all the same species. The closest genetic affiliation of the South Island takahē seems to be to the African swamphen (*Porphyrio madagascariensis*). This idea is supported by the fact that takahē are more similar in colour to that species than they are to pūkeko. The pūkeko, on the other hand, belongs to a very widespread species that is also found across much of Australia.

In terms of timings of events, rails as a group evolved perhaps 40 million years ago. The ancestral rail lineage that led to the gallinules, including takahē and pūkeko, split off around 15 million years ago. It was around this time that some gallinules developed the large frontal shield that is such a defining feature of both takahē and pūkeko. The evolutionary trajectory of the two species split about 4–7 million years ago, and it is thought that the South Island takahē's ancestor settled in New Zealand about 2.5 million years ago. The leaner body shape of the moho suggests it evolved over a shorter time span, so its ancestor probably arrived more recently, perhaps 1–1.5 million years ago. The Australasian swamphen reached Australia about 600,000 years ago, but as mentioned above, it didn't successfully make it to New Zealand until about 500 years ago. In all cases, the dispersing birds had to fly hundreds of kilometres across the Tasman Sea from Australia to reach here.

The living New Zealand cousins

PŪKEKO

Porphyrio melanotus melanotus

It's hard to miss a pūkeko, or New Zealand swamphen, with its bright red bill and frontal shield on its face, its big red feet, and its striking violet-blue neck and chest. The takahē's closest living relative in Aotearoa seems to strut rather than stroll, its black tail flicking

Pūkeko. **ROD MORRIS**

Western weka. **ROD MORRIS**

up to reveal a jaunty white undertail.

If you go back through old editions of field guides to New Zealand birds, you'll find that the pūkeko's scientific name has changed over time from *Porphyrio melanotus* to *P. porphyrio melanotus*, and that recently taxonomists have decided it is actually a subspecies of *P. melanotus*. The species *P. melanotus* is now commonly known as the south-west Pacific swamphen, and it has five living subspecies. These include the Australasian swamphen, *P. melanotus melanotus*, which occurs in Australia and New Zealand. The other subspecies occur in New Guinea and across the Pacific, in Palau, Niue, Samoa, Fiji and Tonga.

WEKA

Gallirallus australis

The weka is a similar size to the pūkeko, reaching around a kilogram in weight. Found only in New Zealand, it is a large flightless rail that is most closely related to the much smaller banded rail (see below). Weka have brown feathers and red eyes, and a bold, inquisitive nature. Four subspecies of weka have been recognised, but the latest thought is that there are really just two forms, the North Island weka and South Island weka. The birds are vulnerable to predation by stoats, and in many places their numbers seem to rise and fall. Altogether, there are eight living species in the genus *Gallirallus*, all found in the Australian and Pacific region; the Lord Howe Island woodhen looks very similar to the New Zealand species.

BANDED RAIL, MOHO PERERŪ

Gallirallus philippensis

Weighing in at 170 grams, the reclusive banded rail was once widespread throughout New Zealand but is now found only in coastal wetlands in the north of the North Island and South Island, as well as on a number of offshore islands, including Manawatāwhi / Three Kings Islands. The banded rail is well camouflaged and cryptic in most of its range, but is much

Banded rail. **ROD MORRIS**

Auckland Island rail. **ROD MORRIS**

bolder and more visible on Great Barrier Island (Aotea Island), perhaps due to the absence of stoats there.

Banded rails fly and are impressive international travellers. The New Zealand banded rail belongs to a very widespread species, also found in South Asia and the Indian and Pacific Oceans.

AUCKLAND ISLAND RAIL

Lewinia muelleri

The most mysterious and isolated of New Zealand's eight living rail species, the Auckland Island rail is found only on subantarctic Adams and Disappointment Islands, which are home to a few hundred birds. It weighs 90 grams and shares its genus with Lewin's rail (*Lewinia pectoralis*) in Australia. Very little is known about the species.

SPOTLESS CRAKE, PŪWETO

Zapornia tabuensis

The 45-gram spotless crake is a secretive swamp dweller. It is widespread across much of New Zealand and on many offshore islands, including Rangitāhua / Kermadec Islands and the Chatham Islands. It is another great traveller, found from the Philippines east across the Pacific to French Polynesia.

MARSH CRAKE, KOTOREKE

Zapornia pusilla

Confirming the intercontinental jet-setting nature of the rails, the marsh crake is the most cosmopolitan of New Zealand's rails, found pretty much everywhere except Antarctica and the Americas. Because of its furtive nature, however, the smallest of our rails – weighing just 35 grams – is rarely seen.

Spotless crake. **B CHUDLEIGH**

Marsh crake. **CRAIG MCKENZIE**

Australian Coot. **GUY VICKERS**

AUSTRALIAN COOT

Fulica atra

You can be forgiven for thinking that the coot is a small duck, as it is most often seen swimming on lakes and ponds with other waterfowl. It is the johnny-come-lately of the New Zealand rail family, having arrived from Australia in the 1950s and rapidly spread through the whole country. Individuals have even been spotted on Stewart Island / Rakiura and the Snares Islands / Tini Heke. The 500-gram bird is also known as the Eurasian coot and occurs in Africa and Asia, as well as Europe and Australasia.

PART 2

REVELATION

1970s and 1980s

Kia kore e rite ki te moa.

Don't let anything else go the way of the moa.

CHAPTER ELEVEN

BIOLOGISTS AT WORK

In 1971, Jim Mills – newly employed in the research section of the Department of Internal Affairs Wildlife Service – found himself spearheading an intensive research programme to find out why the takahē population was in decline and to identify solutions. The new urgency saw a much-needed boost in funding and resources for the Takahē Recovery Programme, which during the previous decade had just been ticking over.

Jim requested the assistance of Roger Lavers, another new employee who was working on mustelids and already had holiday experience working with Brian Reid on takahē surveys in the Murchisons. Field technician Peter Shaw came on board, and Brian remained involved, especially at Pukaha / Mt Bruce, where Colin Roderick was doggedly pursuing the as yet elusive goal of breeding takahē in captivity.

Jim's first priority was to broaden the population study out from the Takahē Valley–Point Burn site. He chose the Miller Peak area in the north of the Murchisons, overlooking the Middle Fiord of Lake Te Anau and its South West Arm, and the Eyles–Wisely area, in the west of the Murchisons. Takahē there lived in the low alpine zone rather than on the subalpine valley floor as they did in Takahē Valley and the Point Burn, and these new survey catchments contained habitat that was much more typical of the Murchisons. Aside from the head basins, much of Takahē Valley and Point Burn are anomalous: the dominant tussock on the valley floor, copper tussock (a subspecies of red tussock), is not common elsewhere in the Murchisons. Later studies would show that copper tussock is much lower in nutrients than other snow tussock species.

Jim's first bout of fieldwork at the beginning of winter in 1971 didn't get off to the most auspicious start: it was two weeks before he got his first glimpse of the bird he was studying, and he wondered what he had let himself in for. But his first sight of a takahē, its bold red bill and striking blue feathers standing out against the clean white of new snow, reassured him that the research was both possible and imperative.

Radio-tracking in the Chester Burn in the mid-1980s. **DAVE CROUCHLEY**

Jim would go on to study takahē for nearly 20 years, often working in the field for six weeks at a time, and he says he never failed to get a thrill seeing these magnificent birds in the dramatic Fiordland landscape: 'they were a beautiful bird in a beautiful area'.[1]

The aim of the study, which would eventually come to an end in 1987, was to observe all the takahē breeding in the three areas and compare how successful they were. It wasn't long before Jim and the team – which importantly included a number of trainee Wildlife Service officers in addition to Roger – were observing nearly a hundred birds across the three study sites. It wasn't easy; the birds had to be banded so they could be recognised, and once they had been handled they became wary of people and actively avoided them. But binoculars and patient observation of the birds as they moved among the all-concealing snow tussocks and shrubs would eventually yield an identification, and data point by hard-won data point, a clearer picture began to emerge.

In the first full year of the study, covering the 1972/73 breeding season, Jim and his team realised that the number of takahē pairs breeding on the valley floors in Takahē Valley and Point Burn had dropped again, down to five pairs. The following year there were just three pairs. But the decline wasn't just on the valley floors – the number of pairs breeding in the low alpine head basins was down to four. In the two valleys, the number of pairs was now half what it was when the birds were rediscovered in 1948.

But it was quite a different story in the other two study areas. In the first year of the study there were 18 pairs in the Eyles–Wisely area and 12 in Miller Peak. Over the following two years these numbers grew, to 21 pairs in the Eyles–Wisely and 21 in Miller Peak.

In addition to counting birds, the researchers weighed the takahē to see how well they were faring in the different areas. The birds in Takahē Valley and Point Burn were around 9 per cent lighter than those in the other areas, and even within Takahē Valley birds from the copper tussock-covered valley floor were lighter than those feeding on other snow tussock species in the head basins.

About 80 per cent of the monitored takahē pairs bred each year, and nearly three-quarters of the eggs they laid hatched. These figures were encouraging, but the news was rather gloomier when it came to chick survival. Three-quarters of the chicks died in the first three months, although if they made it that far they had a slightly better than 50:50 chance of making it to a year old. Around half of the successful nests hatched two chicks, but one of the chicks usually died within a few days and it was a very rare pair that successfully reared two chicks. The researchers found it almost impossible to work out why chicks died. Possibilities ranged from starvation to injury, and predation from falcons, gulls and stoats. The team recorded a notable near-death experience for one Miller Peak chick, when a fast-acting observer rescued it from the hungry bill of a weka. Weka were abundant in the Murchisons up to the 1970s, but in the 1980s their population underwent a dramatic decline of its own and they have been largely absent ever since.

In 1972, Jim invited University of Otago botanist Alan Mark to be involved in the cross-disciplinary research, to better understand what takahē eat and whether deer were competing with the birds; he replaced botanist Colin Burrows, who had a brief involvement in the programme. Alan had been researching *Chionochloa* tussocks for a few years and was an expert in alpine ecosystems, so he was a perfect choice. At first the study remained focused in Takahē Valley, looking at what food takahē preferred to eat from spring to autumn, and at the impact of deer. The research showed that top of the menu for birds in the head basins were three of the five most common species of snow tussock – mid-ribbed, curled and narrow-leaved, in that order – as well as Petrie's mountain daisy. The takahē were very fussy about which bit of the tussock plant they ate – they picked only the most succulent 2–3-centimetre section at the base of each tiller (a tiller is an individual stem, from which sprout several leaves). A popular side dish was the leaf bases of the endemic Takahē Valley speargrass, *Aciphylla takahea*, despite the fiercely protective spike at the tip of each leaf.

If they were available, tussock seeds trumped leaf bases in the popularity stakes. They weren't available every year, but when they were it was easy for the birds to harvest them – they simply stripped the seedheads by running their bill up along the flower stalk. Other plant species made occasional appearances on the menu: a couple of sedges, several speargrasses and a fourth snow tussock species, the round-leaved snow tussock.

Chionochloa tussocks

Snow tussocks are the keystone species of New Zealand's alpine grasslands and herbfields. Although low in stature – never more than waist-high – they are long-lived and structurally as complex as a forest. The genus name *Chionochloa* means 'snow grass', which is another common name for the group, and it contains some 25 species, mostly found in New Zealand. While a handful of the species grow as grassy swards, most have a bunched or tussock habit, with one plant comprising a tight bunch of stems or tillers from which leaves sprout. Three species of snow tussock are of particular importance to takahē in Fiordland: mid-ribbed (*C. pallens* subsp. *cadens*), curled (*C. crassiuscula* subsp. *torta*) and narrow-leaved (*C. rigida* subsp. *amara*). Copper tussock (*C. rubra* subsp. *cuprea*), a subspecies of red tussock, is also a mainstay in some locations.

Botanists now recognise that what they used to call red tussock (*Chionochloa rubra*) in Fiordland is actually the subspecies copper tussock, and what was formerly identified as broad-leaved tussock (*C. flavescens*) is narrow-leaved tussock, which is why many earlier scientific papers on takahē and deer diet refer to these species.

Curled tussock (*Chionochloa crassiuscula* subsp. *torta*) is the second most popular tussock on the Murchisons Mountain takahē menu. JOHN BARKLA

The base of a mid-ribbed tussock plant (*Chionochloa pallens* subs. *cadens*) showing several tillers, from each of which sprouts a few leaves. Takahē are tussock pullers and tiller snippers, pulling out a whole tiller and then nipping off the juicy base. ROWAN HINDMARSH-WALLS

The birds inhabiting the floor of Takahē Valley were living in an environment dominated by copper tussock, and so this obviously featured large in their summer diet. But Alan and Jim were also curious about the birds' winter diet.

Earlier research had shown that in winter takahē leave the snow-covered alpine zone to spend time in the lower subalpine zone, where plentiful shrubs and tall tussocks provide a canopy that effectively shelters the birds from the weather and allows them to rest and feed in a hidden world beneath the snow. If the weather is very bad, they retreat into the lower subalpine beech forest and dig up the rhizomes of *Hypolepis millefolium*, the thousand-leaved fern. This is a member of the bracken family, and is a summer-green fern that loses its leaves when the first winter frosts strike. It grows in light gaps in the forest and among the tussocks, favouring recently disturbed soils such as debris fans left behind by river flooding and old rockslides where the ground is loose and richer in nutrients. A short winter research trip in August 1974, during very bitter weather, revealed that at this time of year the fern rhizomes are very rich in starch.

By 1976, the population research was showing that the survival rate of chicks seemed to depend on the quality of the habitat, with adults in 'good' territories consistently producing chicks, while pairs from 'poor' territories usually failed. The question of how this was related to the quality and quantity of vegetation and whether the quality could be manipulated would become a new important research question.

An expanded food study was set up in new catchments to avoid any further disturbance to the birds in the population study; these included Dana Peaks, Snag Burn and Plateau Creek. At the same time, a new ecologist was brought in to further investigate tussock flowering and relationships between vegetation types and deer diet. William (Bill) Lee had just completed his PhD, with Alan Mark as his supervisor. He had experience in the chemical analysis of plant nutrients and had barely started with the DSIR Botany Division when he was also assigned to help investigate takahē diet. For the next three years Bill found himself working with Jim, Roger and Alan, following takahē to record what they were eating. It was very difficult to observe the birds in the act of feeding, so instead they would find a feeding pair and then follow at a distance recording what was in the trail of discarded plants they left behind. This research went on year-round, including in the frosts and snows of winter, and Bill describes the cold experience as one that ensured he had no further interest in winter sports such as skiing.

The researchers' intensive and time-consuming observations revealed that the birds preferred different kinds of food at different times of the year, and they would hunt these

Bill Lee walking through a large area of *Hypolepis millefolium* in beech forest in the Murchison Mountains in the 1970s. The rest of the understorey was eaten out by deer. WILLIAM LEE

Roger Lavers, Bill Lee and Jim Mills (left to right), in the 1970s, looking at the bright green thousand-leaved fern (*Hypolepis millefolium*) growing in the beech forest amongst taller prickly shield ferns. WILLIAM LEE

out even if they weren't very common. Petrie's mountain daisy was in vogue from October to January, while narrow-leaved snow tussock was mostly eaten over the mating season, in November and December. Curled snow tussock appeared on the menu in spring and autumn, while mid-ribbed snow tussock was the main menu item in midsummer. This got the team wondering if – and how – the nutrient content of these plants changed seasonally.

An initial nutrient analysis looked at the tiller bases of mid-ribbed and narrow-leaved snow tussocks collected from one site in March. It showed that the bases from both species contained the highest concentrations of sugars, amino acids and other nutrients in the plants, and that the takahē's preferred snow tussock, the mid-ribbed, had higher levels than the slightly less preferred narrow-leaved snow tussock. The base was also the juiciest and least fibrous part of a leaf.

The research team then embarked on a month-by-month analysis to find out what seasonal changes were taking place. They tested the three preferred species of tussock, as well as the birds' favourite mountain daisy and *Hypolepis* rhizomes.

This work was carried out alongside an extremely labour-intensive and time-consuming study into how common different plants were across the Murchisons compared to the plants the takahē were eating. Here are a few of the numbers. From October 1976 to June 1977 (early spring to late autumn), the researchers visited three study areas once a month, walking along random straight-line transects through different takahē territories and counting more than 8,000 piles of discarded leaves and tillers to quantify exactly what the birds were eating. To find out how abundant each plant species was, two observers working together surveyed 262 randomly placed plots, each measuring 5 metres by 5 metres. Once a month, the team visited eight different sites to sample five leaves from a number of different plants belonging to different species; these were then analysed in the lab. They snipped and pruned tillers from tussock plants, and then every month measured how quickly they grew back. And they stalked after feeding takahē, so they could sample leaves from a plant that had just been eaten to compare them to leaves from the same species growing nearby that the birds had ignored.

What did this all show? For a start, it confirmed that takahē targeted favoured species. They ate mid-ribbed snow tussock three times as often as you would expect if they were just randomly browsing what was available. Their favourite mountain daisy was targeted five times more frequently than expected. And they browsed curled snow tussock, which was the commonest snow tussock species, half as often as expected, and mostly in October, May and June.

The takahē favoured rapidly growing tussocks, as this was when they contained good levels of nutrients. The discerning diners also knew that mid-ribbed and narrow-leaved snow tussocks reliably had the highest levels of nitrogen, phosphorus, potassium and soluble sugar of all the snow tussocks. When they were laying eggs and feeding young chicks, the birds seasoned this diet with the mountain daisy, which had superior levels of calcium and lipids, as well as good amounts of sodium and magnesium. In the shoulder seasons, when the birds added curled snow tussock to their diet, this didn't have high nitrogen levels but it was blessed with plenty of sulfur, sodium, sugars, starch and lipids.

If the takahē were focusing on a particular species, it wasn't a case of all plants of that species being equal. They were actively seeking out individual plants with higher-than-average levels of different nutrients at different times of the year, such as nitrogen and phosphorus in spring and early summer, prior to egg-laying and chick-rearing, and carbohydrates in late

Project Takahē

In 1981, TVNZ's fledgling Natural History Unit released the first of its award-winning Wild South documentaries: *Project Takahē*. The film-makers – Neil Harraway, Michael Stedman and Peter Hayden – worked with Jim Mills and the Wildlife Service over several years to film a friendly pair of takahē at Plateau Creek. Camera operator Robert Brown captured world-first footage of a pair incubating an egg, feeding on seeding tussock and during winter, when one of the birds memorably sticks its head up and has a lump of snow perched on top. The film also features a helicopter applying fertiliser to various plots in the Murchisons in an effort to create better-quality habitat for the birds.

summer and autumn, when they were building up fat reserves to survive the winter. The birds were very discerning. A takahē would walk straight past most tussocks without even trying them. When it found a plant of interest, it would pull out a single tiller and test it. If the tiller wasn't up to scratch – which happened about a third of the time – it would move on to the next likely plant, and the next one, until it identified a specimen that contained desirable levels of whatever nutrients it was after. The takahē were very moderate in what they ate. They usually pulled out about four tillers, and only in a fifth of cases did they eat seven or more. Even during a heavy feeding bout on a particularly tasty tussock, they never ate more than 10 per cent of the plant's tillers, thereby ensuring its future survival.

This selective feeding seemed to encourage tussock growth, resulting in more tillers. The observation moved Alan to comment that the feeding relationship had obviously developed over time and 'was a clear case of co-evolution between bird and plant'.[2]

A takahē would return over time to repeatedly feed on favoured plants. This was particularly noticeable and most easily observable in the mountain daisy, whose individual leaves could regrow from below the break made by a takahē as it plucked the leaf. The researchers looked at more than 140 daisies takahē had fed on, and found that nearly half of them showed distinctive signs of having been eaten previously.

The problem was that by the late 1970s the takahē no longer had the tussock grassland to themselves. They were sharing it with red deer and research during the decade showed these intruders were eating plenty of the takahē's preferred tussocks, the narrow-leaved and mid-ribbed snow tussocks. The deer were also proving to have as discerning a palate as takahē when it came to picking out narrow-leaved snow tussocks with the highest levels of phosphorus. The deer liked plenty of species that takahē didn't eat, especially large herbs such as the Mt Cook buttercup and Haast's carrot, among others, but it was clear that when deer populations were high they had a devastating effect on alpine tussock grasslands, especially in the wetter western areas of Fiordland. In fact, the deer had already eaten out their favourite species, including the snow tussock hunangamoho, and had shifted their diet to those species favoured by takahē.

Alan had shown in previous research that the problem was how deer browsed. While takahē pulled individual tillers out from the base of the plant, deer munched on mouthfuls of tussock leaves from the top down, cutting them quickly and efficiently with sharp teeth. A hungry deer could eat an entire tussock.

The researchers wanted to quantify exactly what the impact of deer grazing was versus takahē feeding, so they carried out experiments in which they cut the leaves of some mid-ribbed snow tussocks at various heights from the base to mimic deer browse. Other plants got the takahē treatment, with the researchers removing tillers by wrenching them off in a similar manner to the birds. However, the researchers were 'greedier' than the birds themselves, taking a quarter, half and, in some cases, all the leaves from each tussock.

This study plot shows tussocks that were cut to imitate deer browse, and then subsequently devoured by takahē. They took many years to recover. **WILLIAM LEE**

They discovered that deer browsing, from the top down, severely reduced new tiller production. However, extracting tillers following the takahē practice actually stimulated tiller production, increasing overall tussock growth.

In an experiment that took place in spring involving tussocks cut to imitate deer browse, the resident pair of takahē at first ignored the plants. But that suddenly changed in autumn, when the birds began devouring tillers from the cut tussocks – they ate more than 6,700 of them, which was about 10 per cent of the plants. These cut plants were producing lots of starch as they put on new growth to replace what they had lost, and starch is what takahē desire in autumn. Although this feeding frenzy rather interfered with part of the experiment and was a double whammy for the poor tussocks, there were still enough plants left to show that, six months after their trim, those that had been cut were about 40 per cent smaller than control plants that had been left alone. There was also evidence that a number of tussocks had died from the heavy cutting.

The experimental plants were remeasured eight years later, and although the cut mid-ribbed snow tussocks had greater numbers of tillers than the control plants, these contained much less available food than tillers in the controls. The latter, meanwhile, had flourished, expanding in girth and sporting 60 per cent more tillers. This was despite ongoing takahē feeding. It's impressive to consider that the researchers counted every single tiller in their study area every time they returned to measure things, as they reported precisely that at least 3,004 tillers had been eaten.

Nutritional analyses had also revealed why *Hypolepis* was such an important part of the takahē winter diet. The thousand-leaved fern loses its leaves when the first frosts strike, and at this point levels of starch in the rhizomes increase to 20 per cent of the dry weight, with a notable 8 per cent of soluble sugars.

The takahē use their strong beaks to dig up the fern rhizome, which Bill describes as being the takahē equivalent of potato, a sustaining food that supports the birds' winter nomadic existence. For five months of the year it makes up 60–80 per cent of their diet. The birds supplement the *Hypolepis* with leaf bases from hunangamoho, three species of sedge and a rush, which between them have high levels of phosphorus, magnesium, potassium, nitrogen, sodium, soluble sugars and lipids.

While deer didn't eat *Hypolepis*, they did like the takahē's other favourite winter plants, which as a result had been eaten out in some areas of the Murchisons. The researchers were concerned that this put the stress on birds at what is already a nutritionally stressful time of year.

They were also worried that much of the lower storey in the forest had been removed by heavy deer grazing, especially in the drier eastern areas. They speculated this loss of cover would allow the soil to freeze more often and to greater depths in cold winters, making it harder for the takahē to dig out the *Hypolepis* rhizomes. This certainly proved to be the

The omnivorous takahē

Takahē are famously vegetarian – except when they aren't.

While the herbivores are well known for chowing through metres of tussock stem each day, there are plenty of reports of takahē snacking on meat – and not just the invertebrates that their parents feed to them when they're young.

In 2011, a group of schoolkids visiting Wellington's Zealandia sanctuary watched on as famous retiree Puffin nabbed a paradise shelduck chick, and the takahē rangers at Burwood often find decapitated paradise shelduck chicks inside the takahē pens, minus their feet. Out on Tiritiri Matangi, a takahē was seen with a fantail in its mouth. And there is video footage from Sanctuary Mountain Maungatautari of a large chick polishing off a house sparrow. Also on Tiritiri Matangi, there are plenty of observations of takahē down on the beach eating the abundant sandhoppers.

Daryl Eason, who worked at Burwood Takahē Centre, has seen a takahē chick eat a mouse at Burwood, and mouse fur also turns up in takahē droppings in the Murchison Mountains. It's not hard to imagine takahē coming across mice in a beech mast year when the little rodents reach plague proportions. The birds are also very opportunistic when it comes to lizards, which are abundant on predator-free islands, and they happily eat moths, beetles and wētā, as well as worms, cicadas, craneflies and aquatic invertebrates.

case in the winter of 1977, which was exceptionally harsh. In Takahē Valley, temperatures plunged to -17°C, the ice on Lake Orbell was up to 40 centimetres thick, and the metre of snow on the valley floor froze. Inside the forest, which had little of its original understorey remaining, the ground froze to a depth of 10 centimetres. The takahē were unable to dig out *Hypolepis* rhizomes and instead foraged around the edge of the lake and in small seepages, where they grazed indiscriminately on a range of plants. By contrast, takahē living on the upper Snag Burn flats, where half a metre of snow lay on the ground, faced a very different situation in the surrounding forest: the ground remained unfrozen beneath a thick understorey and the birds dug out fern rhizomes as usual.

Having identified the significant impact that deer browsing had on the health of the

Aerial topdressing of fertiliser took place in a few Murchison Mountains takahē territories in the late 1970s and early 1980s to see if increased plant nutrients and growth would benefit takahē. DOC

snow tussocks and the availability of food for takahē, as well as the discovery that takahē selected plants based on their nutritional profile, Jim proposed an experiment to fertilise large areas to see if this encouraged the ecosystem to recover. He was hoping that better-quality food would boost takahē breeding, resulting in fewer infertile eggs and better chick survival. After all, using fertiliser works well with pasture grass and livestock such as cows, and this technique had been previously used to great success in Scotland with heather to improve grouse productivity. It was very difficult to get permission for the work as the area is, of course, part of a national park and there are strict rules governing what can and can't be done there. Modifying an ecosystem is not an acceptable activity.

In late 1976, permission was finally granted for a fertiliser experiment, and the team spread – by hand – different combinations of various fertilisers across 5-square-metre plots of mid-ribbed snow tussock: nitrogen-rich urea, superphosphate as a source of phosphorus, and gypsum to provide calcium. There was a pretty fast visible change. Plants that were fertilised with nitrogen, either on its own or in combination, soon responded by turning a deep greenish colour instead of the usual yellowish green. The small, low-growing blue tussock, which often grows sparsely between the taller snow tussocks, really enjoyed the nutrient boost, and quickly spread to form a continuous lush sward. Within two months, the resident takahē were tucking with gusto into the tussocks that had been boosted with nitrogen or with a super combination of nitrogen, phosphorus and calcium. Based on this success, Jim managed to convince the Fiordland National Park Board to allow some wider-scale aerial topdressing using a helicopter. This finally happened for the first time in October 1979, when pilot Bill Black spread fertiliser across five takahē territories in Chester Burn and Plateau Creek, concentrating on areas of narrow-leaved and mid-ribbed snow tussock. For various reasons it became difficult to follow up the experiments, and despite the results looking initially promising, the pairs in topdressed territories were no more productive after five years than their neighbours in unfertilised areas. It was decided that the end did not justify the means. Besides, deer numbers were dropping rapidly and the team was more hopeful that the ongoing slow recovery of the vegetation would cause a matching improvement in breeding success.

Hunters working on foot had made a significant impact on deer numbers: by the late 1960s, the deer population in the Murchison Mountains was estimated to be just over 2,000 animals, down from about 4,500 in the early 1960s. Helicopter hunting in the Murchisons wasn't allowed because of concerns about disturbing the takahē, but with Jim's firm encouragement it finally got the go-ahead in 1975. As a result, there was a notable increase in the effectiveness of the hunting and deer numbers dropped rapidly – and with this, the large herbs recovered. Roger looked at the stomach contents of deer killed in the mid-1970s and found that they were already eating less snow tussock and had gone back to browsing more of their favourite herbs. The immediate threat of competition by deer for the same tussocks that takahē preferred was averted.

Twenty years after the cutting experiments, Bill Lee returned to resurvey the snow tussock plots. He found that the mid-ribbed tussocks subjected to mock deer browse were still about 30 per cent lighter and smaller than control plants. The researchers estimated that at this rate it would take at least 30 years for the tussock grasslands to recover to their former health and biomass – or possibly even longer, given that mid-ribbed snow tussock is one of the faster-growing *Chionochloa* species and presumably others would take even longer to recover. The recovery, of course, assumes that deer aren't still present in significant numbers. It was clear that Fiordland's alpine grasslands had suffered during the 1950s when deer numbers were at their highest, and the long lag time between destruction and recovery was linked to the ongoing decline in the takahē population.

CHAPTER TWELVE

MANAGEMENT IN THE MURCHISONS

A significant turning point for the Takahē Recovery Programme came in 1978, the 30th anniversary of the rediscovery. The Fiordland National Park Board hosted a special takahē symposium to look back on everything that had been learnt over the past three decades and to plot a new direction for the future. Fifty-five people attended, from a wide range of government departments and non-governmental organisations. Wildlife Service staff from the Department of Internal Affairs were there, including the research section staff, among others. Fiordland National Park rangers were also present, along with members of the park board. Forest Service deer-cullers and researchers were in attendance. DSIR scientists from the organisation's Botany, Ecology and the Remote Sensing Divisions came along, as did academics from several universities. And also at the symposium were representatives from the Ornithological Society of New Zealand, the Royal Forest and Bird Protection Society and the World Wildlife Fund (WWF).

Over two days the attendees listened to talks about the biology of the bird, habitat research, captive breeding efforts, deer control and stoats. There was animated discussion about the impact of deer on the vegetation in Fiordland, and on the newly emerging revelations of the relationship between mast seeding and fruiting years of trees and tussocks and the consequent population explosions of mice, rats and stoats. There was much debate about the use of the toxin 1080 to control rodents and stoats, and its merits and risks, and the pros and cons of using fertiliser to improve habitat. There was also discussion about whether to stay focused on saving takahē in Fiordland versus creating new populations on islands. Dominating the conversation was the simple but alarming fact that takahē were not breeding quickly enough to replace birds that were dying – if nothing was done, the population would inevitably die out.

A number of recommendations came out of the symposium. These were written up and finally approved by the Fauna Protection Advisory Committee in 1982, forming the first management plan for takahē. This was effectively the prototype for threatened species recovery plans in New Zealand, although a formal plan for kākāpō, published in 1989, is officially the first. Such plans have become an integral part of many threatened species recovery programmes, laying out DOC's goals and objectives for the conservation of particular plants or animals.

In essence, the first early takahē plan recommended that the wild population in the Murchisons should be intensively managed to maximise breeding success. The plan approved habitat manipulation – that is, aerial topdressing – to improve habitat quality for the birds. Although most takahē pairs laid two eggs, they generally managed to raise only a single chick, and it was suggested that leaving each wild pair with just one fertile egg might actually increase their chances of success. Excess fertile eggs could then be incubated and hand-reared in captivity; this was a significant change from the captive breeding approach that had been pursued to date at Pukaha / Mt Bruce, and would entail the construction of a dedicated captive-rearing facility. The pipeline of captive-reared chicks would allow a new wild population to be established in Fiordland and birds to be released onto islands that were free of introduced browsers such as deer and predators such as rats and stoats.

Intensive management in the wild kicked off for takahē in 1981, with the first count in what would become an annual breeding season census. Wildlife Service officers wouldn't be appointed as takahē rangers until 1983, so the earliest complete Murchison Mountains surveys continued to be carried out by trainee wildlife officers, as well as the existing research team – including Jim Mills and Roger Lavers, along with Andy Garrick, who had replaced technician Peter Shaw. There was now a well-developed map of all the takahē territories, so the search areas were clearly defined, and there was a significant number of banded birds. As this season was also the beginning of a captive-rearing programme, the searchers were looking for nests containing eggs that were almost fully developed. The idea was to leave each pair of wild birds with one good egg and remove any excess eggs. As it turned out, they didn't find any suitable eggs, and just one fully developed egg was sent to Pukaha / Mt Bruce.

The annual censuses took place on foot, with teams of two helicoptered in to various blocks within the Murchisons for weeks at a time. By now, there was a network of small huts and bivvies scattered across the Murchies, which had provided shelter to deer-cullers in earlier years as well as takahē workers. A portable bivvy, sponsored by the WWF and known as the Panda Bivvy after the organisation's giant panda logo, could be lifted by helicopter from one location to another as needed, and the teams stayed in touch with Arnie Wright at the Te Anau Wildlife Park with a nightly radio schedule. During the day, the teams searched territories looking for takahē sign – droppings and piles of discarded

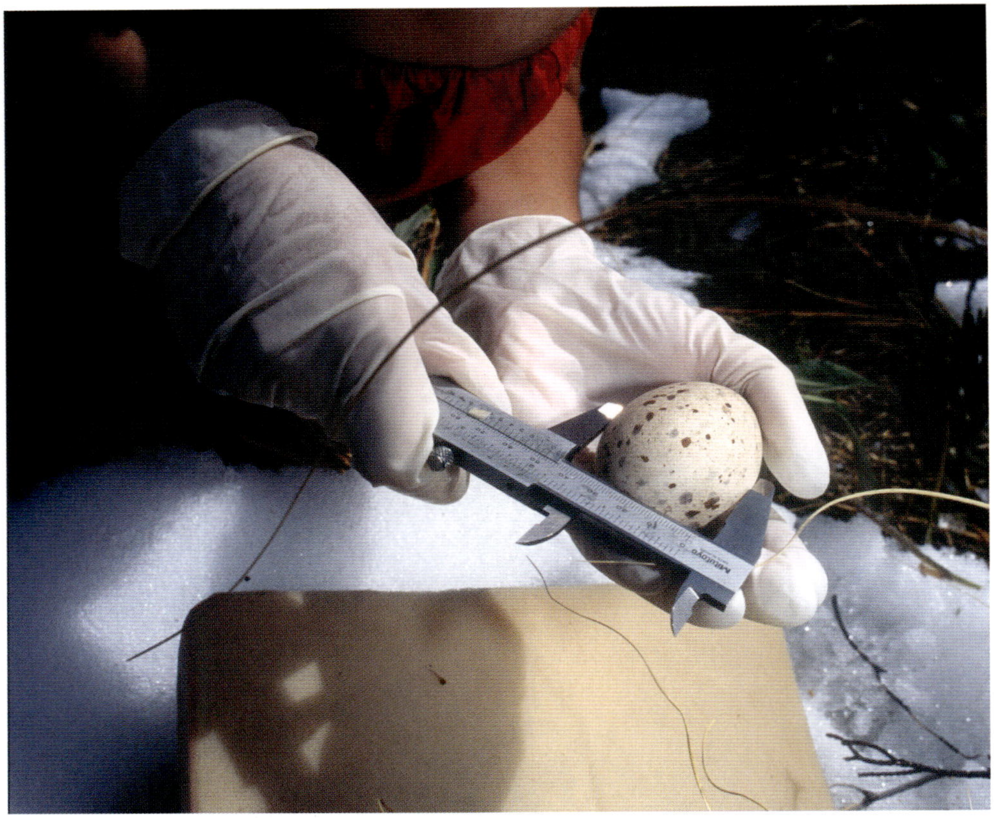

Starting in the 1980s, takahē rangers would try to locate every takahē nest in the Murchison Mountains each year, taking measurements and collecting some of the fertile eggs for captive rearing. DAVE CROUCHLEY

tussock tillers. There was a lot of walking involved, traversing valley floors and sides, clambering into small hanging valleys, and scrambling up rocky guts to gain access to the neighbouring valleys.

When the workers found sign, they would hone in trying to work out where a pair might be nesting, looking for increasing amounts of sign and hopefully coming across a latrine, which would indicate a nest under a nearby tussock. They would move quietly, parting dense tussock leaves and peering underneath, looking for a runway or path that might lead – hopefully – to a nesting bird. Most birds, especially in the eastern valleys, had been exposed to people before and quickly ran away when they were disturbed, but sometimes a diligent parent would remain incubating until the fieldworker reached into the nest. If there were eggs in the nest, the observer carefully candled them to check if they were fertile and to measure their development. First, the observer put their head and both hands inside a black cloth bag that cut out the light. Then, using a small torch that had a soft circle of foam at one end that sat gently against the egg, they shone a light through the egg to reveal the delicate lacework of blood vessels surrounding the embryo inside, from which they could calculate the age of the egg in days. If the egg was infertile, they saw just an undefined mass of yolk. Newly laid eggs cannot be aged as the blood vessels inside have yet to develop.

The drive to put leg bands on as many takahē as possible was still underway, and catching the birds to band them was an energetic process that basically involved running the birds down. The early 1980s saw the first use of dogs and radio transmitters in the Takahē Recovery Programme, and both would help speed up the detection and catching of takahē in the wild immeasurably. Andy led the way with takahē, taking his cue from the Kākāpō Recovery Programme – a breeding population of kākāpō had just been discovered on Stewart Island / Rakiura and dogs were proving vital in the detection of the birds.

In the early days of dog use in conservation, keen owners simply trained the dogs they already owned. In Andy's case, that dog was a Labrador called Van. He was trained to find and retrieve moulting live ducks for research purposes, and to recover shot birds during the game-hunting season. Van had begun working with Andy in the Manawatū in the late 1970s. When it came to takahē, the dog would find a bird by tracking its scent, and then catch it by gently holding it to the ground with his chest or blocking its escape, waiting for Andy to arrive and take the bird off him. Van wore a home-made muzzle while working, and Andy reckons he probably caught about 90 takahē for measurement and marking in the three years they worked together in the Murchisons. Van went on to work on kiwi and kākāpō programmes in other parts of the country. In later years, pointing dogs were more frequently used. These are breeds that point or indicate where a bird is hiding without ever touching it. The first pointer to be used on takahē was Adler, a German short-haired pointer owned by Te Anau-based Wildlife Service fisheries officer Bill Jarvie, and this was followed by Dave Crouchley's dog Tara.

The research team was particularly interested in finding out more about what takahē did in winter, when their subalpine territories were covered in snow. Radio-tracking technology had been rapidly developing through the 1970s, and the transmitters had finally become small enough for large birds such as kākāpō and takahē to carry without too much impact. Scientist Brian Lloyd and others in the Wildlife Service had already created a package that they were successfully using with kākāpō, and Andy was happy to piggyback on this success. These days, a number of companies provide off-the-shelf radio-tracking equipment that is tailor-made for anything from a tiny bird to a full-grown elephant. In the early 1980s, however, you had to configure your own. This involved soldering together the basic transmitter unit and aerial with a small battery and a magnetic reed switch that allowed the tag to be turned off to save the limited battery life when it was being stored. The soldered unit was then encased in a streamlined waterproof shell made from five-minute

Andy Garrick holds a takahē that has been fitted with an early type of radio transmitter. Early 1980s. **ANDY GARRICK**

epoxy Araldite. The package was positioned on the takahē's back and secured there with soft cords, which extended from each corner of the unit. These cords were used to form a strap around the base of each wing, so that the transmitter resembled a miniature backpack.

The radio transmitters emitted a beeping signal on different frequencies that could be detected using a special radio receiver and aerial. The receiver could be bought off the shelf, but the Yagi aerials, as they were known, were another home-made affair. They looked

Wildlife trainee Maurice Wilke with Andy Garrick's conservation dog Van, working in Waterfall Creek. NICK TORR

like miniature television aerials, and were constructed from aluminium tubing that had to be exactly the right length and configuration to receive the signal. This signal allowed the researchers to identify the direction of a bird, and its volume provided an indication of how far away the bird was. The exact position of a bird could be calculated by triangulating fixes from several different locations.

Twenty transmitters were fitted to takahē in the Murchisons in an 18-month period from May 1981 to January 1983, and Andy recalls tracking them through two winters, both on foot and from the air in a helicopter. In the latter, the aerial had to be fitted to the exterior of the helicopter, so Andy rigged up an attachment system using a fibreglass fishing-rod blank tied to one of the skids. He worked with Te Anau helicopter pilot Bill 'Blackie' Black in his Squirrel, and wired Blackie's headphones into the receiver so the pilot could help pinpoint the birds' locations. Andy describes one memorable midwinter trip:

> Blackie and I set off to monitor the transmittered birds after a period of heavy and prolonged snowfall. We flew into the headwater tributary of the McKenzie Burn, which had the hut situated in it, to find the tarn iced over and the entire basin a total white-out. To our dismay, we picked up a radio signal coming from the far side of the tarn in what looked to be

Pilot Richard Hayes tracking takahē from the air. The Hughes 500 helicopter has two radio-tracking aerials mounted at the front of the skids. DAVE CROUCHLEY

very heavy snow cover. We roughly determined the source of the signal, and after circling several times concluded that the best we could hope for would be that the transmitter had come off the bird, but we recognised also that the bird may have perished in the conditions now prevailing. We decided to carry onto the Woodrow Faces and elsewhere to check our other birds, before returning to dig a hole and recover the transmitter with or without a bird attached to it.

In due course, we arrived back and more precisely pinpointed the location of the signal before looking for a place nearby to park the Squirrel. While we were doing that, lo and behold, a blue head and red beak poked up through the snow, followed by another, then yet another, though this one had a dull grey bill! This was the summer territory of a pair of tarks that had reared a chick from the breeding season just gone. The territory was a good kilometre or so from the nearest bit of beech forest, but supported healthy stands of subalpine shrubs, scrub and herbaceous plants, which is where the birds emerged from. On closer inspection, it was evident that although the snow had formed a continuous canopy over the top of this vegetation, underneath were large, almost cavernous, well-insulated air spaces the birds were living in.[1]

At the time, the researchers were happy that this was further evidence that the deer-ravaged vegetation was coming back to health as deer numbers were kept in check, and was capable of supporting takahē year-round. And it was proof that while takahē could spend significant amounts of time in beech forest in winter, they were also capable of surviving long periods of heavy snow, living and feeding in among the shrubs and tall tussocks on the valley floors and lower slopes.

It was a labour-intensive research project. The 20 tagged birds were each tracked for an average of 267 days, coming to a grand total of 5,343 tracking days. None of the birds died from stoat predation, and this was taken as yet further evidence that this threat was not significant.

All this monitoring revealed that, by 1982, the takahē population in the Murchisons had dropped to an alarming 116 birds, a decline of almost 40 per cent in 10 years. There were a further seven birds in captivity and a handful of birds recorded elsewhere in Fiordland – the situation was dire. The following year, Dave Crouchley and Nick Torr were appointed as full-time takahē rangers for the Wildlife Service, responsible for managing the population in the Murchison Mountains.

The intensive Murchison Mountains field research programme that had begun in 1971 wound down as the emphasis shifted to management. Scientist Jim Mills later calculated that over the 12 years of the study, various researchers and wildlife trainees had between them spent 30 person-years in the field. It was no easy task studying elusive birds in a remote wilderness environment, and management would prove to be equally challenging.

Out in the field, meanwhile, newly minted takahē rangers Dave and Nick monitored radio-transmittered birds in the Murchisons, catching them once a year to replace the tags with new ones that had a fresh battery. They had to carry out the annual census too, so a close eye could be kept on the population to see if it was starting to go up or was still declining. This yearly census covered what was considered the core area, and didn't usually include the remote areas further west, where a further 10 or so birds lived.

This field programme sat alongside the captive-rearing programme that was being developed, and a key part of the job involved monitoring nesting takahē in spring and early summer, and carrying out nest manipulation. Research had highlighted that only 30 per cent of takahē chicks that hatched survived their first year of life, and the new plan was to try to improve the number of survivors. Each nesting pair was to be left with a single fertile egg to hatch and rear. One egg would be removed from a two-egg nest and would then either be given to another nest to replace infertile or dead eggs, or taken to the new captive-rearing centre that was being built at Burwood. About 30 per cent of takahē eggs are infertile, and if a nest contained dud eggs and there were no suitable fertile eggs available, it was destroyed in the hope that this would encourage the pair to nest for a second time. In fact, about 40 per cent of these pairs did re-nest.

Where should takahē go?

For several decades following the 1948 rediscovery, there was a vigorous debate between researchers as to the merits of trying to establish new takahē populations outside the Murchisons – that old adage about having your eggs in more than one basket. As early as 1953, when the takahē population was believed to be only 70 or so birds, the naturalist Walter Oliver wrote:

> the takahē is so low in numbers that a catastrophe might adversely affect its chance of survival. Nursing the colony, by proclaiming a colony, will not stay the biological process of extinction ... the time has arrived when ... birds should be transferred to a more congenial climate, protected from vermin, predators and competitors ... and accustomed to different kinds of foods.[2]

By the late 1970s, everyone agreed that, as an insurance policy, it would be a good idea to have a much larger population of takahē spread across several sites – but the spirited differences of opinion lay in where to move the birds. One faction, led by Jim Mills and supported by ecologists familiar with the bird's feeding habits in Fiordland, argued that takahē were subalpine and alpine specialists that would not do well outside this kind of habitat. Another group of ecologists thought that the harsh Fiordland environment might be suboptimal for the birds as it experienced such bad weather. A third group argued that, based on fossil evidence, takahē had once been widespread in the lowlands and were quite plastic in their habits, so would thrive on offshore islands where they would be protected from the risk of predation and competition for food.

Between 1982 and 1989, 104 viable eggs plus four small chicks were removed from the Murchison Mountains and taken to Burwood for captive rearing. The wild population, meanwhile, held its ground, and the new management regime looked promising.

OPPOSITE: Takahē scientist Jim Mills working in Plateau Creek, 1985. PETER MOORE

CHAPTER THIRTEEN

BURWOOD BEGINNINGS

The programme to rear takahē in captivity for release began in 1981, with the intention of testing out a strategy developed by Colin Roderick at Pukaha / Mt Bruce. Based on the previous two decades of experience with takahē, as well as other species, Colin devised a method of rearing chicks using puppets so that the young birds rarely saw humans and would not become imprinted on them, as had occurred with the very first hand-reared takahē in the late 1950s.

Unfortunately, in the first year there were no excess eggs available in the Murchisons that were at the right stage of development for the trial, so it couldn't proceed. However, the field team took the opportunity to measure the temperature in a number of nests during incubation – information that would be useful later for incubating eggs artificially. A single fully developed egg was collected from Fiordland and taken to Pukaha / Mt Bruce, where it was hatched and reared by foster takahē parents. The chick was a male, named Ralph, and in 1989 he would father a single female chick, Terri, with his mate Betty.

The following year, 1982, aviculture expert Martin Bell, who had been working at Pukaha / Mt Bruce, set up in makeshift quarters in a garage at the Te Anau Wildlife Park. The park was home to a Wildlife Service fish hatchery, as well as a small number of rehabilitated birds on display to the public. After a successful practice run with five pūkeko eggs, Martin was ready to receive takahē eggs. In November and again in December, he received two special deliveries from the Murchisons: four eggs at the late embryo stage, each nestled in soft tissue inside a thermos flask warmed to 35°C. They were collected at this late stage because knowledge about incubation was still limited and the facilities were basic. Martin followed the method prescribed by Colin and perfected with the pūkeko eggs, and successfully reared three chicks to breeding age: two males, named Tussock and Snow; and the female Alpine, who would prove to be a pivotal bird in the Takahē Recovery Programme. These birds took up residence at the wildlife park and become quite the tourist drawcard, as

Snow, Tussock and Alpine, seen here as juveniles in 1983, were the first takahē to be hand-reared in Te Anau. NICK TORR

this was still the only place outside Pukaha / Mt Bruce where the public could see takahē.

After the success of the efforts in 1982, Martin concentrated on developing incubation methods so that he could receive wild eggs that were in earlier stages of development. In 1983, he successfully raised six chicks, and in 1984 a further three. The four females and five males were the first takahē to be permanently moved to an island, being sent to Maud Island / Te Hoiere in the Marlborough Sounds in 1984 and 1985 as the nucleus of a new island population (see Chapter 14).

In 1985, the hand-rearing programme moved from the Te Anau garage to a purpose-built facility near Burwood Bush, 40 kilometres east of Te Anau. This was thanks mainly to botanist Alan Mark, who had worked in the Murchisons and was also well known for his high-profile role in the Save the Manapouri campaign in the 1970s, which fought to minimise the extent to which water levels in Lake Te Anau and Lake Manapouri would be raised when the Manapouri hydroelectricity scheme was constructed. Alan had lobbied the Land Settlement Board and the Department of Lands and Survey for many years to protect a large area of undeveloped red tussock grassland at Gorge Hill, part of which runs alongside State Highway 94 between Mossburn and Te Anau (the tussock is now known to be a subspecies of red tussock called copper tussock). He argued that the 1,500 hectares of

The newly built incubation and brooder facility at the Burwood Takahē Centre, mid-1980s.
DAVE CROUCHLEY

tussock grassland was a unique feature of western Southland and deserved to be protected as a scenic reserve, rather than developed. Fellow botanist Peter Johnson suggested that a further 100 hectares that was little modified should also be included.

While a small majority of the local landholders was in favour of the reserve, Lands and Survey seemed intent on continuing to develop and convert tussock to pasture on its farms through frequent burning as well as ploughing. However, the Fauna Protection Advisory Council had been helping the Wildlife Service look for an area of tussock grassland suitable for holding the hand-reared takahē until their release back to the wild, and they recommended that the area be protected. Lands and Survey agreed, and both areas of tussock, as well as an equally large area of mixed beech forest, were designated as a reserve. In addition, a farmhouse and associated buildings on the other side of the road from the Red Tussock Conservation Area were acquired for the captive-rearing facility.

As well as the captive-rearing unit and housing for staff on the site, a 10-hectare area on the tussock–forest margin within the reserve was enclosed with a predator-proof fence to provide a safe habitat for some resident takahē adult pairs and each year's batch of hand-reared chicks. This was soon extended to 80 hectares, and was intended to hold up to six resident breeding pairs as well as the captive-raised juveniles for their first year. The plan was that the resident pairs would raise one chick of their own, and any excess fertile eggs would be included in the maximum of 16 eggs that were hand-raised each year, thus reducing the need to collect so many eggs from the wild. The adults would also foster the hand-reared chicks and, very importantly, teach them how to properly eat tussock and dig for *Hypolepis* rhizomes, and when to seek shelter in the forest during times of heavy snowfall.

Martin was the first aviculturalist at Burwood, where he was joined by Daryl Eason. But Martin soon moved on, and over the next 26 years a number of other aviculturalists spent time at the centre raising takahē, including Ron van Mierlo, Murray Willans, Tarn Fechney, Andrew Smart, Jeremy Carroll and Glen Greaves.

When the field team collected excess eggs in the Murchisons, they packed each one carefully in foam inside a screw-top container, with its air cell facing upwards, and then placed this in a portable incubator. The plan was to collect the eggs and transport them as smoothly as possible to Burwood via helicopter and road within three hours. Once at the facility, they were then transferred to a larger incubator that contained an automatic turner to roll the eggs regularly.

Based on the many eggs that passed through their hands over the next few years, Daryl and Murray calculated that the average takahē egg is 73.5 millimetres long and weighs 96.5 grams when it is freshly laid. Eggs lose weight as the embryo develops and as it loses moisture to the atmosphere. The ideal weight loss is 18 per cent. The key to a successful hatching is to ensure that an egg loses weight at the correct rate, and that depends on the

A takahē chick hatching at the Burwood Takahē Centre. ROSS CURTIS

Dana

Dana lived for only seven years, but she left a lasting impression on Daryl Eason. She was hatched at Burwood in late 1991, from a wild egg collected from the Murchisons. She remained at Burwood as one of the foster parents, and Daryl says she was memorable for her ferocity. He doesn't know why she became so ferocious, and he recounts that the easiest way to deal with her when he went into her pen was to let her attack the daypack he always made sure he was carrying. 'She'd be hanging from the bottom of your daypack with her beak, and you could walk along with her hanging beside you,' he says.[1] However, some people who didn't know this trick ended up climbing a tree to escape Dana's attentions, and she would circle around waiting for them to descend. Despite being so memorable, Dana had only a short stint at Burwood. With her mate Owen, she produced four chicks in four years before dying in early 1999.

humidity it is kept at. Initially, eggs were kept at 37.4°C and 49 per cent relative humidity, and were rolled by 180 degrees every 60 minutes in alternate directions. Under this scenario, weight loss was uncontrolled and only two-thirds of eggs successfully hatched.

Daryl and Murray were determined to improve this percentage and began to refine the incubation method so that they could control an egg's weight loss. With just a few small changes – setting the humidity initially at 52 per cent and turning the eggs every 36 minutes, and then fine-tuning the humidity up and down as required – hatching success soared to 90 per cent. When it was time for a chick to hatch, the egg was transferred to a less humid incubator, where it would be easier for the chick to break through the thin membrane separating the air cell from the rest of the egg. Once the chick had externally pipped (meaning it had managed to make the first small break in the eggshell), it was moved to a very humid hatcher. During the last stages, the chick was played adult contact calls to encourage its efforts. Once hatched, it was then moved to yet another brooder to dry out and gather its strength, at which point it was moved into a brooder with other chicks.

The original hand-rearing technique for chicks, as proposed by Colin and refined initially by Martin, would be incrementally improved over subsequent years, until the following procedure became standard and was used until 2011. It was designed to mimic the care a chick would get from its parents in the wild and avoid the birds imprinting on humans

Very young takahē chicks being fed using a puppet resembling a takahē. The 'mother' model at the right provided warmth and shelter. ROD MORRIS

during the critical first six weeks. After hatching in an incubator, small groups of two to four chicks were kept together in a brooder pen with one-way glass on one side so human observers could keep an eye on proceedings without being seen. The pen had a floor of pasture grass turf, and included planted tussocks and even – like a miniature museum diorama – a painted view of mountains. This enrichment was to encourage the chicks to forage on their own and give them a varied environment. When they weren't being fed during the day, and at night when they slept, the chicks retreated under a model parent suspended over a heat pad, which provided warmth and shelter. To make the brooding as lifelike as possible, a small speaker in the model parent played recorded contact calls.

The chicks were initially fed every 20 minutes during the day, and the intervals were then increased to every 60–90 minutes by the time they were two weeks old. The human feeder began the process by putting two beak-shaped bits of red-painted wood on their index finger and thumb to emulate an upper and lower bill. Next, they pulled on a hand puppet, out of which the beak protruded. The puppet included a long blue sleeve to cover the arm, and contained a small internal speaker which played adult contact calls. The feeder then reached their arm inside the pen and daintily used the 'beak' to feed each chick tiny portions of mash. For young chicks, the mash was a mixture of pureed vegetables and

When they were a few days old, hand-reared takahē chicks were given access to a veranda and outside area, and a variety of foods. ROSS CURTIS

Farex baby food, as well as a more standard aviary chick food. As wild takahē chicks are fed almost exclusively on invertebrates to begin with, the hand-reared chicks also got up to six moths and 12 earthworms each day.

Takahē chicks begin self-feeding quite early, so they were left with a small bowl of mash that they could help themselves to in between their scheduled feeds. They soon cottoned on to these self-serve snacks. Also on the menu was clover, which supplied some essential nutrients missing from the vegetables. And from early on, the chicks received 'grass training' – the feeder would offer stems of young grass, holding them with the puppet beak as the youngsters learnt to snip off sections with their own beak. Over time, the chicks would learn how to pick the grass up with their feet and self-feed takahē style. The chicks were weighed monthly to make sure they were hitting their growth targets.

Once they were 10 days old, the chicks were encouraged to venture out to a small external veranda, where they could experience sunlight and fresh air. They enjoyed bathing in fresh water, and as soon as they were steady on their feet, they were given a shallow pond to splash around in.

At five weeks of age they were introduced to large outdoor pens, and when they were two months old their access to the brooder room was removed. Takahē chicks of this age are still brooded by their parents, so the chicks continued to have access to a model takahē suspended over a nest made from dead grass.

Next, it was time for the chicks to be introduced to some real takahē foster parents. In a kind of foster speed-dating, the adults met up to six foster chicks over a couple of days. After a few more days at the brooder facility to ensure the group was compatible and that the adults were feeding every chick, the group was moved to the foster pair's outdoor enclosure, an area inside the predator-proof fence. The fostering stage was very important for teaching hand-reared chicks how to behave properly. Before fostering was introduced, there had been some behavioural issues such as bullying and feather-plucking, which other birds could have started copying. Fostering soon resolved these problems.

There were always complications and deaths could occur at any stage of the process, but three-quarters of the chicks that hatched survived to a year of age. In comparison, in the Murchisons three-quarters of the wild chicks died soon after hatching.

When the Burwood team first started hand-rearing, the young takahē weren't mentored by older experienced birds when they were released outside but were left to their own devices. Grubbing comes naturally to takahē, but a bird won't learn to dig for *Hypolepis* rhizomes unless it is shown how. Daryl says that in 1989 he began what he termed forced-feeding training. Groups of juveniles were put in a small fenced area containing no food

A takahē pulls out a tussock tiller or grass stem with its strong beak, then transfers it to a foot so it can chew at the juicy base before discarding it. **PETER MOORE**

Daryl Eason playing sound recordings of a female takahē to an egg that is due to hatch in an incubator. **GERALD CUBITT / DOC**

except *Hypolepis*, and he sat with them digging out the rhizomes and feeding them to the birds to give them the idea. They were then left on their own for up to a couple of weeks until they were feeding themselves. It worked, but it was stressful on the birds. The next year, Daryl tested whether an adult bird put in the pen would teach the juveniles, and found that this worked very well. By 1993 there was a new protocol – as long as the younger birds spent time over winter with an experienced adult or pair, they would learn to head into the forest when it snowed and dig for rhizomes. They were not just learning how to feed – they were effectively picking up on social behaviours to help them become a 'proper' takahē able to survive in the Fiordland mountains, which were the destination for many of the early Burwood birds.

A serious issue that occurred in the first decade of the hand-rearing programme at Burwood was dietary deficiency – most likely a shortage of manganese – that caused a number of chicks to develop dislocated patella, or kneecaps, when they were a few weeks old. As a result of this condition, the chicks couldn't bend their legs or walk properly. Thirteen chicks, or 10 per cent of the 135 that were puppet reared between 1985 and 1994,

The Burwood Takahē Centre after a heavy midwinter snow fall. SAM HAULTAIN

were affected, and all but one had to be euthanised. The deficiency was easily solved by adding a simple potassium permanganate solution to the chicks' daily feed, and from 1995 onwards there were no further dislocated kneecaps.

One bird, Houdini, was operated on and had his dislocated kneecap tied back. He limped badly as a result. Houdini was the son of ferocious Dana and Owen. He never mated or produced chicks, but he was partnered with another male, Tom, and the two of them remained at Burwood, where they became excellent foster parents.

While a few birds such as Tom and Houdini remained at Burwood as breeding or foster pairs, most of the early cohorts of captive-reared birds were destined for release either on islands or in the Stuart Mountains.

Ins and outs of a takahē's diet

How much food does a takahē eat?

What nutrients do takahē get from their diet? Is mid-ribbed snow tussock, which dominates the western Murchison Mountains, better than copper tussock in the eastern Murchies as a source of nutrients? And if they were to be moved to islands, could the birds survive on grass instead of tussock? In 1984, technician Peter Moore was tasked with answering these questions – and in the process he would also find a definitive answer to the question of how much poo a single takahē produces in a day.

Peter had access to the first three birds to be hand-reared at the Te Anau Wildlife Park: Alpine, Tussock and Snow. He knew them from the colour of their plastic leg bands as the female Red, and the males Yellow (Tussock) and White (Snow). He set up an initial feeding trial in 1984, although he had to cancel it after three weeks as a heavy snowfall made it impossible to collect tussock.

In February 1985, the experiment began for real. The birds were housed in a garage, each in a 1 metre by 1 metre stainless-steel cage with a mesh floor. The birds were given time to acclimatise to their surroundings and their managed diet before the experimental regime was introduced.

Peter roped in a team of Wildlife Service trainees and wage workers to assist – basically, anyone who was around. Every day, they dug up six large copper tussocks (which were known then as red tussocks) from a nearby farm, collected armloads of mid-ribbed snow tussock along the side of the Borland Saddle Road or near Homer Tunnel on the Milford Road, and picked bags of cocksfoot grass from fields near the wildlife park. All this was taken to the park for time-consuming processing. This involved stripping the tough outer leaf sheaths, and cutting the tussock from the base into uniform 15-centimetre lengths and the cocksfoot into 30-centimetre lengths. Peter needed 4.5 kilograms of each standardised food each day, which equated to 1,000–2,000 neatly trimmed tillers of each variety. It was a lot of work – and the team had to repeat the process more than 50 times.

After the acclimatisation period, the experiment ran for three replicates of 11 days, making a total of 33 days. Each bird was fed each food type for 11 days, seven of which were to settle the birds and four of which involved intense scrutiny. The day started at 6.30 am with a weigh-in, and the birds were fed as much as they wanted until 9 pm. Everything that went into the cage was measured, and so was everything that came out – the bits of stalk that weren't eaten, and all the droppings. A takahē, it transpired, eats about 1.3–1.8 kilograms of food a day – about half its own body weight – and excretes about 1 kilogram of waste.

Ruth Kissel collecting mid-ribbed snow tussock for the takahē nutrition experiment (top) and the team turning thousands of tillers into standardised lengths. **PETER MOORE**

The top row shows two snow tussock and one grass species cut to size. In the middle are stems eaten and discarded, and at the bottom are the resulting takahē droppings. **PETER MOORE**

At the end of the experiment, Peter rewarded the birds with a two-day feast of insects, grains and poultry pellets. He reports that the takahē were so content with this all-you-can-eat buffet that they stayed in their cages even when the doors were left open.[2]

How much poo does a takahē do?

The takahē digestive system is a marvel of prolific inefficiency. A takahē spends 90 per cent of daylight hours eating, which can be anything from 10 to 19 hours a day depending on the season. Its diet is mostly tussock in Fiordland, or pasture grasses on the islands, and it eats just the soft bases of tussock tillers and the bases and lower stems of grass blades. All those tussock bases and grass stems pass quickly through the bird's short gut, which is just 1.4–1.9 metres long and acts mainly as a juice extractor. After a short transit lasting just a few hours, the stems are excreted almost unchanged, in roughly 10-centimetre-long cigar-shaped droppings that look like miniature hay bales.

Peter Moore collecting takahē droppings and discarded stems - everything that went in and out of each cage was measured precisly. **PETER MOORE**

Right from the early days, it was clear that a single takahē leaves a lot of sign behind, and several researchers were determined to put an exact number to that. Back in the early 1970s, Brian Reid carried out 24-hour observations of two captive takahē at Pukaha / Mt Bruce that were fed chook pellets and given free access to pasture grass. He measured a daily poo trail of more than 5 metres, with in excess of 4 metres of that coming from grass. He estimated that a takahē on a grass diet alone would produce more than 8 metres of droppings.

Peter Moore's 1985 nutrition experiment offered a more precise measurement for a bird eating tussock. His eyebrow-raising statistic was that a takahē produces droppings at the rate of one every 10 minutes, totalling a mighty trail of 9 metres of poo each day, weighing about 1 kilogram.

A couple of times a day, a takahē also deposits a black tarry dropping, which originates in a part of the gut known as the caeca. All birds except parrots have a pair of caeca, which are finger-like sacs, opening off the junction of the small and large intestine. These structures are involved in nitrogen cycling and the fermentation of carbohydrates, and also help in water retention. Food seems to spend more time in the caeca and is more thoroughly digested there than in its rapid transit system through the intestine.

CHAPTER FOURTEEN

TO THE ISLANDS

Apart from a brief, unproductive visit to Kapiti Island for a couple of Pukaha / Mt Bruce birds in the early 1970s, takahē weren't moved to offshore islands until the mid-1980s. The idea of establishing new populations on islands that were free of introduced browsers and predators had slowly gathered momentum after the 1978 takahē symposium. Apparently, the Fauna Protection Advisory Committee came within one vote of agreeing to transfer all takahē out of Fiordland to alternative locations, either at Pukaha / Mt Bruce or on islands. This ambitious and controversial plan was scaled back to six pairs and, in the end, bad weather stopped it from proceeding. However, once captive rearing was up and running, producing chicks that needed a home, islands were suddenly back in the picture.

The first takahē to head to an island were sent to Maud Island / Te Hoiere in the Marlborough Sounds. The island had been farmed, but in the 1970s Forest and Bird fundraised to buy it from its owners to act as a nature reserve, and the Wildlife Service and Department of Lands and Survey made up the balance. Bush regeneration was underway and a few kākāpō had already been moved there when it was decided that the island's grassy pastures might be a good place for takahē to graze. The first hand-reared birds from the initial trials at the Te Anau Wildlife Park headed off in 1984 and 1985: four females – Mrs Blue, Mrs Black, Mrs White and Maud; and five males – Mr White, Tane, Mr Black, Mr Blue and Bruce. In 1987, the four-year-old Mr and Mrs White successfully raised Maud Island's first takahē chick, a male called Pipi.

In 1988, six young adult takahē from Te Anau and Pukaha / Mt Bruce were moved to Mana Island, off the coast from Porirua, north of Wellington. Mana is a former farm and quarantine station with plenty of remaining pasture grass, although revegetation is well underway in places and over time the number of takahē territories has declined. The same year Rima was the first Mana Island chick to hatch; her mother was two-year-old Squeak

Squeak – super-breeder

Squeak showed early breeding prowess, producing her first chick, Rima, at just two years of age, and she would go on to become legendary in a genetic sense. In 1989, she paired up with Taku, after both birds were moved to Kapiti Island. These two would go on to establish a dynasty of 11 offspring, single-handedly creating a genetic bottleneck by passing on their genes to many descendants in the small island populations.

To compound matters, after Taku's death in 2000 Squeak produced a further five chicks, making a grand total of 17 offspring, 10 of which would go on to produce 62 grandchicks between them. In her post-Taku years, Squeak's love life got very confusing. She was part of a threesome with Bargie and Aho Ake (her grandson), and either male could have fathered her chicks. At one point she was partnered with her own son, Montague, and produced a chick, Ihi, that was simultaneously her daughter and granddaughter. Squeak died at the very respectable age of 19 in 2006 on Kapiti Island, where she had spent most of her life. With her five mates, she successfully produced chicks in 13 breeding seasons between 1988 and 2005.

and her father was six-year-old Tussock. Rima would go on to live on Mana Island until her death in 2000 from gout.

Rima's parents had important Te Anau connections. Tussock had hatched from one of the three eggs collected from the Murchisons in the first successful hand-rearing year at Te Anau, while Squeak was the offspring of the other two birds from that year, Alpine and Snow. Among the other founder birds on Mana was a cohort of captive-bred birds from Pukaha / Mt Bruce, including Taku, who hatched in 1984.

In 1989 Kapiti Island, a well-forested island lying just to the north of Mana Island, received its first takahē, including Squeak and her fellow Mana birds, who were temporarily relocated while a mouse eradication project was underway on Mana. Later the same year, Kapiti's first chicks were produced, including Iti, Squeak and Taku's first offspring, and Terri, whose parents were Betty and Ralph. The latter pair were originally from Pukaha / Mt Bruce – Betty was raised by her mother Red-Left and an unnamed male, while Ralph hailed from the Murchisons but had been hand-raised at Pukaha / Mt Bruce.

Mataku – modest breeder

While some takahē such as Squeak are super breeders, others produce hardly any offspring. Squeak's original partner, Tussock, re-paired with a female raised at Pukaha / Mt Bruce called Mataku, and together they produced two chicks. Mataku later paired up with Snow (Squeak's father, who was moved to Mana Island from Te Anau) and produced three more chicks. But although five is a respectable number of chicks, two of those never bred and Mataku ended up as grandmother to only five grandchicks, a reminder of how uneven takahē breeding effort can be.

This was the beginning of the practice of moving birds between islands. In 1991, two male birds from Maud Island headed to the Hauraki Gulf, to Tiritiri Matangi Island: Mr Blue, one of the original birds moved to Maud; and Stormy, who was born on Maud. Tiritiri Matangi was another former farm in the process of being revegetated, but it differed from the other takahē island homes (which were all island reserves and received very few visitors) in that it was an open sanctuary on the doorstep of New Zealand's largest city. It was already becoming an increasingly popular day-trip destination thanks to a regular ferry service, and the arrival of the two male takahē introduced the species to a large and very thrilled audience.

University of Auckland ecology professor John Craig and botany professor Neil Mitchell had long been agitating to move takahē somewhere public like Tiritiri. They were the masterminds behind the Tiritiri Matangi Island restoration project, which was widely seen as a pipe dream back in the early 1970s when the pair and their students began to look at the ecology of the island. The island had become a recreation reserve in the Hauraki Gulf Maritime Park in 1970, and in 1982 a working plan for its restoration was adopted. Hundreds of thousands of trees were planted in the rank pasture grass and the population of introduced Pacific rats was eradicated. Although much of the island was destined to become forest, significant areas were left as grassland to ensure sites of early Māori occupation remained accessible and undamaged. It was these areas that would suit takahē.

John had studied pūkeko for many years, and more than anyone else he saw the pūkeko-like attributes that takahē still possess. He dismissed Jim Mills' belief that takahē were alpine and subalpine specialists, and thought they would easily adapt to pasture grass growing in a lowland setting. He recognised that no single island would be large enough

Takahē and the historic lighthouse on Tiritiri Matangi Island. MARTIN SANDERS

to hold a self-sustaining population of takahē, but suggested the idea of a metapopulation, whereby takahē were treated as a single, albeit widely dispersed, population that should be managed by regularly moving birds between locations to break up family groups and ensure continued gene flow.

In light of all of this, the first takahē translocation to Tiritiri Matangi Island – involving two males – wasn't a serious proposition but rather a trial. Mr Blue and Stormy had a steady relationship, built nests and brooded an artificial egg, so the following year they were given a fertile egg from Maud Island to see if they could successfully raise it. They were diligent foster parents, although their foster daughter Matangi survived only 11 months before dying from an infection. Because the boys had settled well in their new home, the next takahē to come their way, in 1992, was Matangi's older sister, JJ, who had been hatched and raised

A young takahē sprints along a path. Tiritiri Matangi Island. **MARTIN SANDERS**

by her parents, Becka and Maud, on Maud Island. JJ took up with Stormy, and in January 1993 they hatched Aroha, the first of many takahē conceived on Tiritiri Matangi Island.

By 1991, there were eight pairs breeding on Maud, Mana, Kapiti and Tiritiri Matangi. All up, between 1987 and 1993, 24 founder takahē were moved to and between the four islands.

Island translocation has become a key tool in New Zealand conservation, but the takahē island translocations stand out for a significant reason. While most species are moved between similar habitats – for example, from one forested habitat to another – takahē have been moved between very different places, from high altitudes to sea level, and from a low alpine and subalpine tussock shrubland to ecosystems dominated by exotic pasture grasses and regenerating forest. The translocations have been highly successful, but it wasn't all plain sailing.

OPPOSITE:
European pasture grasses substitute for snow tussocks on islands such as Tiritiri Matangi. **MARTIN SANDERS**

CHAPTER FIFTEEN

TO THE STUART MOUNTAINS

Jim Mills and Roger Lavers have long been proponents of the idea that the best place for takahē is subalpine and alpine Fiordland, and they were keen to establish a second population there. Years before intensive takahē management and the first island transfers began, they were collaborating with various ecologists and with a DSIR remote-sensing expert to find other sites in Fiordland that best matched the Murchisons in terms of having the necessary food species and suitable nearby forest where the birds could retreat in winter. Over a period of three years they assessed nearly 30,000 hectares, concluding that 40 per cent of northern Fiordland could be suitable, while southern Fiordland was much less suitable. They particularly liked the look of the Stuart Mountains, just to the north of the Murchison Mountains. The advantage of the adjacent Stuart Mountains was that, as they expanded, the two populations could merge. In the meantime, the ranges were sufficiently isolated that the two populations would remain quarantined from each other. This was important, as the source birds for the Stuart Mountains would come from the newly established breeding centre at Burwood and there was some anxiety about disease risk from the captive birds. If, for any reason, a disease was introduced to the Stuart Mountains, it was imperative it didn't reach the Murchisons.

For this reason, the first takahē translocations were all one-way journeys: eggs and chicks could come out of the Murchisons to Burwood, but nothing could return there. And birds could move from Burwood to an island or to the Stuart Mountains, but they couldn't come back.

However, the Stuart Mountains had an initial political drawback: the New Zealand Deerstalkers Association wanted this area managed specifically for wapiti, which were prized trophy heads. Wapiti had been introduced from North America in 1905 specifically to form a hunting herd (see Chapter 8), which was confined to the Stuart Mountains and centred around the Glaisnock Wilderness Area. The wilderness designation meant that no

The Stuart Mountains, seen from the Miller Peak area in the northern Murchison Mountains, across the Middle Fiord of Lake Te Anau. SOUTHERLY STORM PHOTOGRAPHY

huts or tracks could be constructed and no aircraft were allowed to land. Unintentionally, it also meant that no species could be introduced – not even rare natives.

The hunters recognised that deer would be in direct conflict with takahē for food, and indirectly in that hunters themselves would likely be excluded from an area containing takahē, just as they are excluded from the Murchisons. The initial recommendation to the National Parks Board and the Fiordland National Park Board to introduce takahē to the Stuart Mountains was declined, and it took much debate and representations to various parliamentary committees before the desired agreement was reached. The Fiordland National Park Board was then able to remove the wilderness status from a small part of the Glaisnock–Edith area of the Stuart Mountains, making a takahē translocation possible.

Between 1987 and 1992, 58 young captive-reared birds from Burwood were released in the Glaisnock area, in groups of two to eight birds at a time. A quarter wore radio transmitters so their survival could be tracked, and every year a census was carried out in catchments immediately around the release site.

The area was difficult to work in, being more remote than the Murchisons and lacking that area's good network of tracks, huts and bivvies. And, unfortunately, the radio transmitters and their harnesses proved very unreliable, making tracking birds very challenging.

A yearling takahē fitted with a radio transmitter. JANE MAXWELL / DOC

Although fieldworkers managed to confirm that about a quarter of the birds survived their first year and only 7 per cent definitely died, that still left more than two-thirds of the birds simply unaccounted for.

Two breeding pairs were confirmed, and one of those made several breeding attempts in two different years. Overall, however, the translocation didn't seem to be succeeding as hoped, and in 1993 a decision was made to stop any further releases. A thorough survey in the summer of 1994/95 could locate only eight birds, seven captive-bred birds and one chick born there. This was a survival rate of less than 14 per cent, much less than the 60 percent survival rate for captive birds released into the Murchisons. Early signs from Maud and Mana Islands, where captive takahē had also been released, also seemed much more promising in terms of survival, at more than 70 per cent, and breeding, with 28 chicks hatched.

There were plenty of reasons why the translocation didn't really work. Deer had arrived in the Stuart Mountains earlier than in the Murchisons and there was less deer control here, so the vegetation may have been in a worse state than predicted, making it harder for the takahē to find high-quality food. Also, half the birds released had never grubbed up and eaten *Hypolepis*, and this important takahē winter survival skill isn't innate knowledge – young birds have to learn it by watching older ones. Not knowing the technique would

Do any takahē survive in the Stuart Mountains?

Throughout the 1990s rangers continued to check parts of the Stuart Mountains, reporting a handful of birds each time. But many of them were single, and in 2002 two of these solo birds were relocated to the Murchisons to give them a better chance of finding mates.

Despite ongoing takahē sightings reported by trampers, hunters and helicopter pilots, there s no good evidence that any takahē still remain in the Stuart Mountains. The takahē team would be delighted to find unknown birds for the genetic potential they could bring to the population, but on summarising a recent search there, Glen Greaves of the Takahē Recovery Programme quipped, 'We went in search of takahē and didn't find anything except goose poo.'[1]

have made surviving winter much more challenging. Three-quarters of the released birds simply hadn't spent any time with adult birds and lacked basic takahē social and life skills.

The youngsters were only a year old on release. They hadn't yet formed pairs and there were no wild experienced birds for them to learn from. With only small numbers of birds, the chance that an individual looking for a mate would bump into one was pretty slim. So, the birds wandered far and wide in their searches for mates, and although the Stuart Mountains form a peninsula, there was nothing to stop the takahē wandering off into the vastness of Fiordland. The original idea had been to release 80 birds, and certainly overseas research has shown that an effective founding population for many species is 80–120 birds, so it may also have been the case that too few birds were introduced too slowly.

Another reason for abandoning the Stuart Mountains releases was that the Murchison Mountains population decreased markedly in the early 1990s, and the decision was made to start returning Burwood birds there to top up the main wild population.

New Zealand's graveyard of birds

New Zealand used to be a place like no other, with a rich, diverse and unusual fauna. This young and geologically restless land, with its mild oceanic climate, was an isolated evolutionary laboratory. Once part of the great southern continent of Gondwana, New Zealand – or more properly, the partially sunken continent of Zealandia – came into being 80–65 million years ago. Biologically, it was home to a few ancient Gondwanan species that came with the land and many more species that dispersed here – and still do so – across a widening Tasman Sea.

The result was a terrestrial biota rich in distinctive lineages found nowhere else. On the animal front, these include tuatara, leiopelmatid frogs, skinks and geckos, land snails and a rich array of land birds and seabirds. It is an unbalanced fauna, though, and there are many things missing – in particular, there are no terrestrial mammals, bar a handful of small bats.

Gigantism, flightlessness and longevity are hallmarks of many of New Zealand's animals. There were the nine species of megaherbivore moa, weighing up to 250 kilograms. Giant flightless geese roamed the grasslands, while large, predatory adzebills stalked the shrublands. Four of the world's flightless perching birds lived here, out of just five known. There is the flightless nocturnal kākāpō, the heaviest parrot in the world. And, of course, there are the takahē and their now extinct cousins the moho, yet more flightless giant vegetarians and the largest members of the rail family.

Then, around 1270 CE, a cataclysm of unprecedented proportions began in New Zealand: the first Polynesian settlers arrived, bringing hunting skills and tools, along with kiore (Pacific rats) and kurī (Polynesian dogs). Within just 300 years, 27 bird species were extinct, including all the moa, adzebills and Haast's eagle. Small birds, especially their eggs and chicks, were vulnerable to predation by kiore. Others, like the moa and takahē, were large, flightless and easily hunted. Adding to this, forest clearance began early and was wide-scale.

The arrival of Europeans from the late 1700s, and the subsequent introduction of many more exotic species, unleashed a new wave of destruction. A quarter of the original 223 breeding bird species in the New Zealand region are now extinct. The distinctive, ancient endemic bird species have been hardest hit: at least 41 per cent have been extirpated. And if there were a cemetery in New Zealand to mark extinct species, then the section for rails and their relatives would, sadly, be well populated with headstones.

The lost rails

Some 750 years ago, when early Polynesians settled here, the New Zealand region boasted perhaps 19 breeding species of rail, not counting vagrant individuals of a further three Australian species that are occasionally recorded here.

By the time of European arrival, several rail species were already extinct, and by the end of the 1800s that tally had risen to 11. The species that lost the battle for survival were the endemics. Most of the surviving species are native and are widespread elsewhere.

The mightiest of the lost rails were the adzebills. The first adzebill bone discovered was so big that it was mistakenly thought to have come from a moa. But as more bones were discovered, it became clearer that this was no relative of the moa.

There were two species of the flightless adzebill, or *Aptornis*: the North Island adzebill and the South Island adzebill. They were – by far – the largest Gruiformes species ever to exist. The larger South Island adzebill weighed in at 19 kilograms, compared with 16 kilograms for the North Island species. Think about that in terms of the takahē – which at a mere 3 kilograms is the remaining heavyweight of the rails – and it is easier to appreciate the bulk of this bird.

What's more, the adzebills were no gentle vegetarians. They had only vestigial wings but they did have strong legs for scratching through soil and leaf litter, and a massive adze-like bill that the famous palaeontologist Sir Richard Owen thought the birds would use 'for frequent, deep thrusts into the ground, and that the quest was for animal, not vegetable matter'.[2] All the evidence, including recent research using stable isotopes, supports the idea that adzebills were predators, and that lizards, tuatara, burrow-nesting birds and large insects were probably all on the menu.

Exactly where the adzebills belonged on the avian family tree and where they stood ecologically were long-standing enigmas that have only just been solved. Taxonomists had long put the adzebills in the Gruiformes, an order that used to include rails and cranes, as well as an assorted grab-bag of species that didn't belong anywhere else, such as New Caledonia's flightless kagu. For a while, the kagu was considered the closest relative of the adzebills, a relationship thought possible because a land bridge once existed between New Zealand and New Caledonia.

New genetic evidence, however, shows that the adzebills are most closely related to several groups of birds from Africa and Madagascar: the flufftails and the Madagascan wood rails. What is particularly striking about this relationship is that the flufftails are very small birds, weighing just 25–50 grams. And the Madagascan wood rails, while larger at up to 280 grams, are also significantly smaller than their extinct New Zealand cousins.

As strange as this geographical connection seems, there are other examples of links between New Zealand and Madagascar. It was recently shown that the closest relative of kiwi was the enormous and extinct Madagascan elephant birds, while New Zealand teals have genetic links to the Madagascar teal. So how did the ancestors of adzebills, kiwi and teals reach far-away Aotearoa? It used to be thought that they were ancient Gondwanan species that moved with the land when the supercontinent split up. But molecular evidence indicates that kiwi and adzebills almost certainly flew here, reaching New Zealand long after it split off from Australia. The flying adzebill ancestor, which would have been a much smaller bird than its descendants (perhaps the size of a wood rail), probably travelled here via Antarctica – which was in a very different place to where it is now, although the bird would still have had to cross open ocean.

If the ancestral adzebill was about the size of a wood rail, then consider how much it grew in size once it lost the ability to fly – it increased in size more than 50-fold. But islands, it seems, are good at growing giants (see feature on page 211).

The adzebills separated from their closest relatives about 40 million years ago. We don't know exactly when the ancestor of the adzebills reached New Zealand, but fossils of another adzebill species found at St Bathans in inland Otago date back 16–19 million years. This extinct species was already flightless and already a giant, which suggests it had probably been here a long time and had survived the Oligocene submergence, a period around 25 million years ago when much of New Zealand – including the North Island – was underwater.

It seems likely that only a single species of adzebill survived the sinking of much of New Zealand, in the South Island. Then, around 1.5 million years ago, an isthmus of land formed across the Manawatū Strait, which had separated the South Island from the newly reappeared North Island following the Oligocene submergence. The land bridge lasted a million years until modern-day Cook Strait formed. It seems that the adzebill was able to move into the North Island, and over time it evolved into a new species. The molecular evidence suggests this happened about a million years ago.

Adzebills were once widespread across both islands. However, their bones have been found in early Māori middens, and scientists believe they were hunted to extinction around 1400 CE, within 200 years of Polynesian settlement.

In addition to the adzebills, there was a suite of smaller, flightless rails. What unites these birds is their small size and long beaks. The snipe-rail, like kiwi and snipe, used its long bill to probe soil in search of invertebrates such as worms. At just 240 grams, it was too small for Māori to bother hunting, but kiore, which arrived with the Polynesian settlers, were more than happy to eat the species to extinction.

The snipe-rail was gone by the time Europeans arrived, but three more small rails survived in the Chatham Islands until the late 1800s. The Chatham Island rail was a diminutive 60 grams, while Dieffenbach's rail weighed about 340 grams, and its tameness was its undoing. At a meaty 2 kilograms, Hawkin's rail probably filled the niche taken by weka on the mainland. Like most rails across the Pacific, the three Chatham rails were likely descendants of the banded rail.

Artist's impression of the extinct North Island adzebill. **PAUL MARTINSON / TE PAPA**

The Australian territory of Macquarie Island is biologically – if not politically – a New Zealand outpost. It too, had its own unique rail, which was last seen just one year after it was described in 1879.

Also on the extinction roll call were two coots – one each for mainland New Zealand and the Chathams – and the dainty Hodgens' waterhen.

The 11th lost rail was, of course, the moho or North Island takahē. Evolutionarily speaking, it had left its pūkeko-like ancestor well behind and was a hefty 4 kilograms, although it was still more slender than its cousin to the south. Even so, its size made it an appealing meal, and apart from a single tantalising record from 1894 in the Ruahine Ranges, it is thought to have died out as a result of human hunting before Europeans arrived.

PART 3

ENDEAVOUR

1990s and 2000s

Mō tātou, ā, mō ngā uri ā muri ake nei.

For all of us and the generations that follow.

CHAPTER SIXTEEN

ANATOMY OF A MAST YEAR

It's late spring, and I'm walking in southern Fiordland. The forest is a mixture of tall podocarps such as rimu and miro interspersed with plenty of silver beech, which is becoming more abundant as the track climbs higher. My attention is mostly on my feet and the track, on the tangles of roots, the mossy carpet and the scatter of debris that has rained down from above. In among the latticework of broken sticks and browning leaves on the sides of the path, I begin to see something I can't quite place – a gentle drift of beige texture that softens the ground like some sort of botanical fog. When I kneel down for a closer look, I realise that it is a carpet of spent flowers, tiny clusters of petals resembling bran flakes – thousands of them. I've never seen them before, but I realise these are silver beech flowers. I imagine that if you had flown above the forest in the weeks prior, the usually green canopy would have been ablaze with red flowers. Now it will be full of freshly germinated seeds that will develop over the coming months to ripen and fall in autumn.

The next day I see some beech flowers for the first time. They are on stunted shrub-sized mountain beech trees, at the bushline where the forest shrinks into the subalpine zone. It's colder up here, so the trees are later to bloom. The flowers are a rich ruby red, borne in luscious clusters that remind me of berries. I've never seen beech trees bedecked with such floral jewellery – the effect is stunning. Nearby are other potent visual signposts of the coming nationwide mega-mast event – snow tussocks surrounded by haloes of flowering spikes, each stalk hung with bleached blonde baubles like a grassy Christmas tree. There are gifts aplenty for the local forest inhabitants.

****

Clouds of pollen rise from the forest during a southern beech mast flowering. ROD MORRIS

There are years in Fiordland when the normally dull-green beech forest blushes an intense red in spring, as millions of tiny flowers synchronously burst into bloom. When the wind blows, massive clouds of pollen billow above the forest canopy, ensuring plenty of flowers will be pollinated. These events occur when the previous summer was hotter than average, and the one before that was cool. That sequence has primed the forest trees to respond in a mass breeding event known as a mast year. These occur every two to six years and have profound consequences for the inhabitants of the forest.

Over the next few months, the pollinated flowers develop into small seeds weighing just a few milligrams each, which ripen and fall in late summer and autumn. The beech forest mast has delivered a seed bonanza whose abundance overwhelms the forest animals – especially the invertebrates, which in a less fruitful year might eat all the seeds. The seeds are simply so numerous that many will escape being eaten and survive to germinate and grow.

It's not only southern beech trees that have mast years – many New Zealand plants do. The irregular breeding of the kākāpō, for example, is tied to the occasional mass fruiting of podocarp trees such as rimu.

Snow tussocks in the alpine zone also respond to warm summers with irregular mass-flowering and mass-seeding years, but there are a number of snow tussock species in the

DeltaT model

Scientists have come up with a simple, elegant mathematical model to describe the cycle of mast seeding, which is known as the DeltaT model. The letter 'T' stands for temperature and delta means 'difference'. If the difference in the average summer temperature from year one (cool) to year two (warmer than average) is big enough, then year three will be a mast year. Scientists used to think that the temperature needed to reach a certain level for a mast year to be triggered, but they've since realised that it doesn't matter what the exact temperatures are – what's important is the relative difference between years. The beauty of the DeltaT model lies in its predictive powers, which in recent years have allowed conservation managers to anticipate coming mast events and plan wide-scale suitable predator control as part of DOC's nationwide Tiakina Ngā Manu – Battle for Our Birds programme.

Murchisons and not all flower at the same time. A snow tussock mast is sometimes in synchrony with the lower-altitude forests and sometimes not.

Mast-flowering and mast-seeding events in beech forest are great news for native animals living in the forest and alpine zones. There is an exuberant abundance of food that enables everything from the lowliest caterpillar to the largest parrots to feast and breed. Species such as orange-fronted parakeets and kākā, which fail to breed most years, successfully raise full clutches of chicks.

But these days there is a dark lining to this silver cloud. In the low alpine and subalpine zones mice populations surge during a tussock mast event, while in beech mast years introduced mice, ship rats and stoats thrive in the forest. The resulting rodent and mustelid plagues wreak havoc on the native wildlife.

An estimated 50 million beech seeds fall across each hectare of forest in a big mast year, or up to 15,000 seeds in a single square metre. It's a veritable smorgasbord for small invertebrates, which make the most of both fallen flowers and seeds, and for mice and rats, which gorge on both seeds and on invertebrates. A single mouse can scoff more than 1,000 protein- and energy-rich seeds a day, and there are bountiful caterpillars, spiders, wētā and beetles to add to its diet.

Mouse plague

Wildlife Service employee Roger Lavers recollects that during a mouse plague in the Murchison Mountains sometime in the 1970s, the subalpine tussock grassland was alive with the rodents, day and night. His colleague Jim Mills also remembers 'the grass moving' with so many mice. Roger invented a mouse trap that comprised a large kerosene tin filled with water, with a piece of No. 8 wire stretched across the top. The wire was threaded through an empty milk powder tin that had been coated in honey and then rolled oats. Mice clambered across the wire to reach the lure, but when they attempted to climb onto the tin it would spin, dropping them into the water. The takahē team would leave the trap in a hut, and on their return find a bucket of mouse corpses; Jim remembers once catching 35 mice in a single night.[1]

Mice numbers rise exponentially during tussock and beech mast years, as there are so many seeds available to eat. **NGA MANU IMAGES**

With unlimited food comes unrestrained breeding, and the prolific rodents produce back-to-back litters throughout winter and the following spring. A female mouse might breed up to 10 times a year and give birth to litters of up to 14 babies, while a female rat can produce up to 10 young every two months. All of those offspring can start breeding when they are just a few weeks old, so numbers add up very quickly. While normally there might be one or two rats in a hectare of forest, one study in northern Fiordland recorded a terrifying population density of 17 rats per hectare during a large mast event, while mice can be as abundant as 50 per hectare. It's no exaggeration to say that there are rats and mice everywhere, running around even in broad daylight.

But the good times never last. As winter progresses, the rain of seeds comes to a halt and the plentiful piles on the forest floor begin to disappear as seeds germinate or rot. A year after the beech forest flowered, the hungry times begin. Over the next six to eight months, a large population of increasingly desperate rats begin to hunt anything they can find – including mice and small rats, along with nesting birds, eggs and chicks. The stoat population spikes in spring as well-fed females give birth to large litters of well-nourished pups. Again, these have no problem finding food and surviving – for a while, at least.

As the second year of the mast unfolds, and when stoat numbers are at their highest, the mice and rats disappear. The starving stoats become bolder in what they will attack, and their target is now local birds and large insects. But it's not enough to sustain the population, which undergoes a dramatic crash in numbers. Two years after it began, the effects of the beech mast finally grind to a close.

A beech mast is a complex web of successive booms and busts, and the picture is made even more complicated as we're not talking about a single species of beech. There are five types found across New Zealand, three of which are important in central Fiordland: red, silver and mountain beech. Red beech is the largest beech species and is found in warmer, more sheltered sites such as valley floors, on rich soils. Mid-sized silver beech grows on mountain slopes on good soils, while the smallest species, mountain beech, is also the hardiest, tolerating poor, dry soils. The species are quite different from one another and have their own reaction to the subtleties of temperature, so they don't always synchronise – at a site with all three species, it's possible to get a single species masting one year or all three masting simultaneously. An important feature of red beech when it comes to its impact in a mast year is that its seeds are twice the size of those of the other two species and therefore have a much higher energy density.

Beech masts don't occur every year and are usually quite localised, affecting just 10 per cent or so of New Zealand's beech forests at a time. But every few years there is an outstanding event known as a mega-mast, when up to half of the country's beech forests experience a simultaneous mast. The years 1979, 1982, 1995, 1999, 2008, 2014, 2016 and 2019 were all mega-mast years (these were the years when seed fell in autumn – flowering

Peter Moore installing a seedfall tray to measure the quantity of beech seed falling each year.
PETER MOORE

was in spring during the previous calendar year). In between, there were much smaller mast events scattered across New Zealand. The Murchison Mountains, with its local climate and own mix of beech species, experienced the big years and in addition has had a few smaller masts of its own.

CHAPTER SEVENTEEN

STOAT THREAT

I've seen plenty of stoats – at a distance. They are usually racing across a road during a mast-seeding year, when their numbers are high. I once watched one in the middle of the day, out on the runway at Invercargill Airport. It was gambolling by itself in a frenzy of jumping, spinning and tail-chasing. This 'weasel war dance', as it's known, is surmised to be a mustelid's attempt to mesmerise prey such as rabbits. I found it comical and chilling at the same time. There were no rabbits nearby, so I couldn't tell if the dance was having its desired effect.

I've also met one stoat at close quarters and it truly lived up to its reputation of being fearless despite its small size. This particular animal was being held in captivity at Lincoln University, where researchers were trying to discover better methods of catching and killing the predators. The stoat was resting in a plastic tube, and when I got close enough to peer into the end of the tube nearest the fence, it immediately twisted round to face me, hurled itself at the wire mesh and snarled a guttural spit of fury and contempt in my face. 'Natural born killer' is indeed a very apt description of the mustelid.

****

When takahē were rediscovered in 1948, there were immediate concerns about the possible impact of stoats on the birds, but it would take nearly 60 years to find out just how devastating these predators can be.

Stoats (known as short-tailed weasels in North America) were brought to New Zealand from Britain in an attempt to control rabbits, which had been introduced here in 1852 to provide food for early European settlers. By the 1870s rabbit numbers had reached plague proportions and they were causing major damage on farms, competing with sheep for grass.

Stoats are capable of killing birds much larger than themselves, such as this kererū. **ROD MORRIS**

It was thought that mustelids, the rabbit's natural enemies in its native Europe, should be introduced to control their numbers, and so in the late 1870s the first shipments of ferrets reached New Zealand. This was followed by introductions of stoats and weasels. In just 10 years, beginning in 1883, some 7,838 stoats and weasels were released. They spread throughout the North and South Islands within 30 years, and while ferrets remained most common in open grassland areas, stoats in particular became widespread throughout all New Zealand ecosystems, including subalpine and alpine zones. They are good climbers, so they are quite at home in forest environments.

It soon became apparent that all three of these fearless killers almost immediately switched their attention to native birds, in addition to the rabbits they had been brought in to eradicate. New Zealand's birds had evolved to avoid avian predators that hunt by sight during the day, such as falcons, but were oblivious to the dangers of mammalian predators, which hunt by smell, often at night.

A stoat is the perfect killer. It is small, weighing no more than 250 grams and measuring less than 40 centimetres from nose to tail tip, but it is strong and fearless and will take on animals many times its own size. Stoats don't live long, but they are precocious and start breeding when they are very young – males first breed when they are just 10–11 months old. Female stoats are able to finely tune how many young they produce, depending on how many prey there are. They evolved this ability to cope with unpredictable irruptions of their prey species in Europe, which unfortunately means they are well suited to coping with beech mast years and rodent outbreaks in New Zealand. A female stoat gives birth in spring, and before their eyes open any females in the litter will already have been impregnated by an adult male. When they leave the nest about a month before the summer solstice, the precocious young females are carrying fertilised embryos whose development is paused at the blastocyst stage. This condition is known as embryonic diapause or suspended implantation, and happens in more than 130 mammalian species ranging in size from mice to bears.

On average, a female stoat produces eight to 10 embryos, but is capable of having up to 18 in a good year with plenty of food. The blastocysts implant in the uterus in late winter and develop for another month before being born. The implantation stage is when the female decides how many embryos to implant – if it is a poor food year, no embryos might implant as she marshals limited food resources to ensure her own survival. She can also limit how many embryos survive to term, and how many nestlings survive.

Stoats eat a range of small animals, including insects, small birds and eggs, but their main food in New Zealand forests is mice and, to a much smaller extent, rats. Research led by biologist Caroline King during the 1970s revealed that in a peak mouse year following a mass-seeding event of southern beech or snow tussock, female stoats will produce large litters that are born when mouse numbers are at their highest.

But mouse outbreaks during a mast-seeding year are short-lived. Once the seed supply runs out it isn't long before mice numbers start to drop, and hungry growing stoats then turn their attention to anything else they can feed on. Researchers would eventually learn that 'anything' might be an animal as large as a 2.8-kilogram takahē. As mice numbers crash, many young stoats starve and die, and those that survive won't produce any young the following summer.

Another area in which stoats excel is travel – young stoats disperse over great distances. The record holder was a young female being radio-tracked in the Eglinton Valley, which

Takahē versus stoat

Jim Mills and Roger Lavers report that Les Stanley was working for the Forest Service in the Woodrow Burn area in the late 1960s or early 1970s when he heard a disturbance and went to investigate. 'He found a stoat with its mouth around the neck of a takahē. The takahē was walking around with the stoat attached. Les was able to approach and hit the stoat with a stick, which detached the stoat and the takahē ran away.'[1]

Takahē don't always need human help to get rid of an attacking stoat. During a later study of nesting takahē that used remote trail cameras to monitor activities at various nests, fieldworker Jane Maxwell reported footage of a takahē shaking off the stoat that had attacked it.

In the early 1970s, Roger and Jim came across another surprising incident:

> We were in the Point Burn trying to determine whether the pair of takahē still had chicks. We came across the pair in an area of dense, tall red tussock approximately 15 metres away. They had not seen us, and so we sat down and waited for over an hour hoping that the pair would move so that we could see them. After more than an hour they had not moved, so we went to investigate. In the area where they were feeding, we found an adult stoat with its head cleanly cut from the body as if it were cut with scissors. No other injuries were evident. The blood on the severed neck and the head was just starting to congeal. There could be no explanation other than a takahē using its beak to cut the head from the neck. [After all] the beak is very strong and can cut thick basal portions of tussocks cleanly.[2]

lies past Te Anau on the road to Milford Sound / Piopiotahi. She disappeared from the area where she was caught and was then trapped a couple of weeks later at Burwood Bush, more than 60 kilometres away in a straight line.

Stoats and the possible threat they pose was certainly front of mind for the first takahē researchers, as they recorded the unexplained disappearance of numerous chicks. In one of the first scientific reports on the newly rediscovered birds, Robert Falla wrote: 'Naturally, predator pressure is to be feared, but so far direct evidence of it is lacking, and the traces of stoat feeding activity examined have consisted of remains of small passerine birds and mice.'[3] Nonetheless, to be on the safe side some stoat traps were put in Takahē Valley and

A stoat killed in a DOC200 trap. SAM HAULTAIN

Point Burn within a few months of the rediscovery. But although the traps caught a few stoats, they were often not set or checked due to concerns over takahē disturbance.

In the first two decades after the 1948 rediscovery, 14 dead adult takahē were found in Takahē Valley and Point Burn. Stoats were positively identified as responsible for three of those deaths, two of which occurred in the 1966/67 breeding season, when a nest was also predated and a chick killed. In many cases no firm cause of death could be identified.

In late 1972, researchers began an intensive stoat study in the Murchisons, with 140 Fenn kill traps set along transects in Takahē Valley and Point Burn. They ran the traplines for four years, before these were scaled back to a smaller number of traps. The researchers also noted when and where they saw stoat tracks in the snow, and they mapped stoat scats and collected them to identify what the animals were eating. Mice, insects and small birds were the main menu items. The researchers caught 27 stoats in nine months in Takahē Valley and Point Burn, but commented that they continued to find just as much stoat scat after the stoat trapping as before, suggesting the trapping was not as effective as had been hoped.

Between November 1976 and May 1977, seven banded takahē disappeared suddenly from a study population of 22 birds and an eighth was found dead. On the other hand, in several alpine valleys where there was very little sign of stoats being present, no takahē died. At the same time, a stoat research project that was trapping in the nearby Eglinton Valley was catching much higher numbers of the predators than usual, following a large southern beech mast the previous summer.

At the 1978 takahē symposium (see Chapter 12), researchers expressed their concerns about stoats as takahē predators and suggested carrying out targeted trapping during the takahē breeding season following a beech mast season. However, within a few years they had reassessed this correlation and decided that the decline in takahē numbers at this time was likely not caused by stoats. Because it was so hard to collect unequivocal evidence for stoats causing takahē deaths, the emphasis for management and research continued to be on maintaining habitat quality through deer control rather than predator control. It was thought that a limited and labour-intensive trapping programme during a stoat outbreak would succeed only in killing animals that, given the high mortality rate of young stoats, would probably have died naturally anyway. In addition, reports of adult takahē fighting off an attacking stoat had provided some reassurance that the adult birds were too large to be killed by a much smaller mustelid.

Twenty years later, the takahē management team were still saying that 'stoats are known to prey upon takahe, but their impact on the Fiordland population is unclear'.[4] Takahē Recovery Programme ranger Jane Maxwell reported at the 50th anniversary takahē symposium in 1998 that, over an eight-year period, the bodies of 51 dead radio-tagged takahē were recovered and of these 2–10 per cent had been killed by stoats, all within 20 months of a beech mast year. The impact of this predation was downplayed, however, as all the birds were non-breeders.

CHAPTER EIGHTEEN

CONSOLIDATION AND COLD WINTERS

In 1987, the Department of Conservation was formed, combining the Wildlife Service (which had traditionally looked after rare animals, along with freshwater fisheries), the Department of Lands and Survey (responsible for national parks and reserves) and some of the Forest Service (which managed wild animal control). This meant that the takahē team, formerly part of the Wildlife Service, no longer had to juggle the sometimes contradictory requirements of different aspects of takahē management, and negotiate actions from other government departments with different priorities. Instead, everything – from threatened species conservation to deer control and national park management – was under one roof.

Around this time, some long-standing staff left the takahē programme and over the next few years others joined. Jim Mills and Roger Lavers moved on, as did Nick Torr. Dave Crouchley stayed until 1990, when he left to work on Maud Island for a few years and looked after the resident takahē there. He then returned to the takahē programme in the mid-1990s. Jane Maxwell, who had started out with the Wildlife Service in 1985, became a long-time fieldworker and would go on to do a Master of Science degree on takahē. University of Otago zoology professor Ian Jamieson, who had worked with pūkeko, became Jane's master's supervisor and also became involved in the Takahē Recovery Group.

The 1990s were a time of consolidation, questioning and waiting for the members of the Takahē Recovery Programme. There was plenty of research underway, conducted through species management and designed to ensure that the goal of growing a healthy takahē population remained paramount.

Throughout the 1990s and into the 2000s, the Burwood takahē-rearing centre reliably produced eight to 16 chicks each year. Between 1991 and 1997, 78 of these captive-reared birds were released into the Murchisons. Excess fertile eggs continued to be collected from wild pairs in the Murchison Mountains, and depending on the year and circumstances at the time, young wild chicks were also often collected. All were then hand-reared. The five to six resident breeding pairs living in tussock pens at Burwood also produced eggs, some of which were collected to be hand-raised, while some were brought up by their parents. Over the years, the team at Burwood had worked out that chicks did better the more time they spent with adult birds rather than with puppet substitutes. So, when they were just a few days old, all the chicks were rotated between hand-rearing and foster parents, spending as much time as they could with the latter in small pens at the back of the rearing unit. The male–male pair of Houdini and Tom (see Chapter 13) were exceptionally reliable foster parents and trainers for many years. Fostered chicks were more relaxed and less aggressive towards other chicks, and learnt more quickly how to hold grass tillers in their feet and to feed themselves from a bowl of mash.

But it wasn't all plain sailing. Burwood had various issues with egg and chick production during the 1990s – and that turned out to be something that was happening in the Murchison Mountains. The Burwood takahē team experienced a number of years when the takahē embryos weren't properly positioned in the egg, the rates of embryo death were higher than expected, chicks needed help to hatch, and chick navels (where they had been attached to the yolk sac) didn't heal properly. Initially, the team suspected that the incubators weren't working exactly as required – and sure enough, when they carried out trials using turkey eggs, which were an easily available substitute, they found that the ageing incubators had unreliable temperatures and rolled the eggs a little roughly. These problems were easily fixed in 1997 with new incubators.

Despite the incubator faults, however, most of the turkey eggs actually hatched successfully, which made the team wonder whether some of the problem lay in a change of quality in the takahē eggs. It turned out that what was happening through the early to mid-1990s to cause that change in egg quality was a series of exceptionally cold winters in Fiordland.

In the early 1990s, Jane Maxwell began research for her master's degree, investigating the survival rate of captive- and wild-reared takahē chicks in the wild. This involved spending many months – summer and winter – over a number of years radio-tracking juvenile birds in the Murchison Mountains. Despite winter cold, Jane would still find takahē out and about foraging for food in the subalpine zone, often on the shady south-facing slopes where the best areas of *Hypolepis* were to be found – those 'takahē potatoes', as Jane and others fondly refer to the rhizomes. At other times, birds were in the forest foraging for *Hypolepis*.

One winter trip into Takahē Valley is etched in Jane's memory. She and partner Simon Torr arrived in the valley to find it was -18°C, and she remembers lighting the fire in Takahē Valley Hut in an effort to melt a large accumulation of ice at the back of the fireplace where a leak had frozen. Ironically, she recalls, the chimney was notoriously bad at drawing, so they had to have the hut door open to let the smoke out – meaning they didn't get much

Many lakes in the Murchison Mountains freeze in winter. Top McKenzie hut is visible at the far right. NICK TORR

benefit from the fire. The next morning, the pair walked up the middle of Lake Orbell, which was covered in thick ice and would remain frozen that year for about three months. The lowest temperature recorded in the valley that winter was -22°C.

This was just one of a series of very cold winters – of the seven winters between 1991 and 1997, five had mean temperatures that were lower than anything recorded since local Murchison Mountains weather records began in the early 1970s. During the 1980s, the mean winter temperature was usually around zero degrees; in 1991 it was -1.6°C, and in 1992 it was -2.6°C. These two winters were cooler than average around the world following the eruption of the Mt Pinatubo volcano in the Philippines, which sent vast quantities of sulphur dioxide into the stratosphere, creating a hazy aerosol that blocked out sunlight. A couple of warmer winters followed, but then there was a run of brutal winters.

At the end of the cold 1992 winter, thick snow lay on the ground for much longer than usual. Jane recalls visiting a territory in the Dana Peaks–Snag Burn area at the time when takahē are beginning to think about breeding. She watched the resident pair make their way through the late snow to high in the subalpine zone, moving purposefully and prospecting around the general area where they usually nested. After a while, they retreated to lower-altitude forest, defeated by the snow. When Jane went down to the forest, she noticed that the *Hypolepis* ferns were already putting out shoots and drawing down on the sugars stored in their rhizomes, making these a much less energy-dense food for the takahē. She realised that the birds had been looking for access to the tussocks they would normally switch to eating in spring, but had been unable to reach them. Jane says it was a tough year for young takahē – most had managed to survive the bitter winter, but they were now facing starvation. Four of the wild-reared takahē yearlings she had been following died that spring. Three of them had lost 40 per cent of their body weight and had almost certainly starved to death; the fourth body was too decomposed to draw any conclusions.

Captive-reared takahē chicks fared better as they were kept at Burwood for their first winter, and although there were sub-zero temperatures and snow at the centre, conditions were not as extreme as in the Murchisons. The young birds were released into the mountains in spring, when they were nearly a year old. The following winter would be the real test of their survival skills.

Times would get tougher for the Fiordland takahē before they got better. In 1995, the average winter temperature in Takahē Valley was a bone-chilling -4.5°C. This was a memorably cold year across the whole country, with Ophir in Central Otago hitting -21.6°C, New Zealand's second-lowest recorded temperature. These exceptionally low temperatures followed a heavy snowfall, and in places sheep were frozen stuck to the icy snow. The following winter in the Murchisons was barely any better – the mean temperature was -4.0°C

The impact of these winters was dramatic. It turns out that the size of the takahē

Jane Maxwell checking the weather station in Takahē Valley. JANE MAXWELL / DOC

population in the Murchisons very closely tracks the mean low winter temperature – if the winter is warmer than average, numbers rise; if the temperature drops, so does the number of birds. The Murchison Mountains takahē population plummeted from 145 birds in 1990 to just 99 birds in 1992, following a cold winter with an average temperature of -2.5°C. This was even fewer birds than during the critical period of the early 1980s, when intensive management began, and it saw the end of captive-reared birds being released into the Stuart Mountains. Instead, they were released back into the Murchisons, to try to boost numbers there.

Cold winters impact takahē in several way. First, for every 1°C difference below average winter temperature, the chance of a takahē dying increases by 20 per cent. This is partly because staying warm is a challenge, and also – as Jane noted – because finding food becomes harder. The takahē struggle to grub enough *Hypolepis* rhizomes out of the frozen ground, and heavy frosts kill young plants that takahē rely on for food. Big snow years are even deadlier. Each extra metre of snow increases a bird's chance of dying by 50 per cent. Deep snow blankets the vegetation, making it more difficult for the birds to access food, and deadly avalanches are more frequent.

The effects of a cold winter in the Murchisons reach well beyond the mountains themselves. The takahē hand-rearing team at Burwood discovered the same relationship between mean winter temperatures and egg survival – a warm winter means a good hatch rate, while a cold winter has the opposite effect. A cold winter compromises both the quality and quantity of eggs laid, probably because the adults have lost weight and condition. In such seasons, fewer pairs nest, fewer eggs are laid in each clutch, and eggs are more likely to be infertile and, very significantly, have embryos that are in the wrong position.

A big research question in the 1990s was whether the Burwood programme of captive-rearing takahē for release back into the wild was worthwhile. Over the decade, 96 captive-raised birds were released into the Murchisons, but did these birds survive as well as wild-reared birds? More importantly, did the captive birds have the same success at pairing and breeding? And were captive-reared birds equally successful on islands and in Fiordland?

In Fiordland, the answer initially seemed to be yes. The first few years of data collected from following birds wearing radio transmitters showed that captive-reared takahē survived as well in the Murchisons as their wild cousins, even given poor survival during harsh winters for both sets of birds. The Burwood birds were no more or less likely to fall off a bluff, drown in a river or die from starvation or cold than the Murchison birds. But survival for the Burwood birds was clearly better if they paired up with a wild bird early on, rather than another captive-reared bird – a bird raised in the wild had useful life skills to share with its partner. And more Burwood females survived because they generally found a wild Fiordland mate, whereas Burwood males tended to find females that had also been captive reared.

A group of yearling takahē at Burwood, acclimatising to winter cold. ROSS CURTIS

The question of whether Burwood birds were good breeders was more difficult to assess and took longer to answer. Breeding success improves over time, and so a few years are needed for birds to gain age and experience. Also, sample sizes early on were small and it was hard to show statistically significant differences, so the better performance of wild birds was noted but largely ignored.

Out on the islands, the situation was slightly different, as every island bird either hailed from captivity or was descended from a captive-raised bird. By the year 2000, there were 59 takahē on the four original islands of Maud, Mana, Kapiti and Tiritiri Matangi. Numbers had climbed steadily, despite the island birds producing fewer juvenile takahē per egg laid than Fiordland birds. The island birds laid lots of infertile eggs and many eggs failed to hatch, although on the plus side a takahē chick that made it to a month old was more likely to survive to breeding age on an island. Although the island populations weren't growing as quickly as originally hoped, they were certainly fulfilling their role of back-up insurance populations and the takahē team was already looking for new sites for future takahē populations.

The takahē population wasn't the only population on the increase in the 1990s – deer numbers were also on the rise. The deer population in the Murchison Mountains had dropped from about 4,500 animals in the early 1960s to about 300–350 animals by the late 1980s, following sustained helicopter and ground hunting driven by a strong demand for wild venison and, later, live deer capture for the booming deer farming industry. But deer values dropped by the early 1980s and so did the commercial hunting effort, and by 1997 their numbers had risen to an estimated 500. DOC then started a renewed effort to keep deer numbers down, and by the end of the decade twice as many deer were being shot and caught, and population numbers had been brought back down to about 400.

A target population of 350 deer was set, as it was thought this level would allow the quality of the vegetation to be maintained and prevent too much competition with takahē for food. To achieve this would require culling 120 deer a year, so a more rigorous deer control programme was established. This was managed by Dave Crouchley and used a combination of hunters shooting from helicopters and trapping deer in live capture pens to supply animals to deer farms. By the end of the decade, deer numbers in the Murchisons were at their lowest, and the takahē population had recovered from the series of cold Fiordland winters and levelled out at 221 birds across all sites.

How takahē die

To put it simply, takahē are accident prone, especially in rugged Fiordland. It's impossible to put a figure on it, but it's unlikely that many takahē die a quiet death from old age.

The list below is of known or probable causes of death cited for takahē in the Murchison Mountains in 2011–12. These were usually for birds carrying radio transmitters whose bodies were found, although often these are too decomposed for the Massey University vets who carry out the necropsies to identify how the individual died.

- misadventure
- leg caught in a bush
- fell through ice and drowned
- myocardial fibrosis (heavily scarred heart)
- predation
- presumed predation
- found submerged in stream and probably drowned.

Two deaths were listed as having unknown causes, while one bird is thought to have actually died from old age.

The following year, four birds drowned during heavy flooding. It was also a stoat plague year, and 15 birds wearing radio transmitters died; four deaths were definitely predation and the balance was almost certainly predation. In other years, takahē have also died in rockfalls and avalanches, and from starvation.

A falcon scavenging a young takahē. **ROD MORRIS**

CHAPTER NINETEEN

LIFE ON THE ISLANDS

Takahē soon settled down to life on the islands. From just eight breeding pairs of takahē on the four islands of Maud, Mana, Kapiti and Tiritiri Matangi in 1991, there were 21 breeding groups by 2000, involving pairs, threesomes and even a foursome. However, the island birds had been left to their own devices, which had resulted in many pairings between close family members. Now, to mix things up socially and genetically, there were to be new introductions from Burwood and birds also were variously moved between islands.

In the early 1990s, researchers carried out an interesting experiment to try to boost the number of chicks each takahē pair produced in a season. The conservation world was agog at the recent successful cross-fostering of highly endangered Chatham Islands black robin chicks by tomtits, their close relative, to increase the number of chicks reared. In this, first clutches of robin eggs were fostered to tomtits so the robins would nest again, effectively doubling the number of eggs produced. To minimise disease risk, the Takahē Recovery Programme wasn't allowing island takahē to return to either Burwood or Fiordland, so Judah Bunin (a student of University of Otago zoologist Ian Jamieson) decided to attempt cross-fostering on Mana Island with what was on hand: the pūkeko, a close relative of the takahē.

By September 1993, six years after the introduction of takahē on Mana, there were five breeding pairs and a trio, along with a rapidly growing population of pūkeko that had colonised the island naturally. The experiment used eggs from these birds, plus eggs brought from Kapiti and Maud Islands, all fostered by Mana's pūkeko. Takahē eggs are quite a bit larger than pūkeko eggs, so to make sure the pūkeko would accept the imposter, the birds were first given a large dummy egg. In total, 12 eggs were fostered over two breeding seasons. Eight of these hatched and two survived their first year, at which point observations stopped. Two out of 12 doesn't sound much, but it was no better or worse than the success rate of eggs left with takahē parents. And just to keep things interesting,

The male Bargie mating with his partner Ihi, on Kapiti Island. GEOFFREY DE LISLE

one takahē pair ended up successfully cross-fostering a pūkeko chick, which was last seen back with a large group of pūkeko. This was not the first time that this had happened: takahē had successfully hatched and reared pūkeko chicks during earlier fostering experiments with captive birds at Pukaha / Mt Bruce.

Cross-fostering the first takahē chick was quite the group effort. The chick – named Mana – was fed by a breeding group of four adult pūkeko, as well as the two juveniles from their first nest of the season. Mana's biological parents were Squeak and Taku, who had been moved from Mana to Kapiti. She remained with her extended pūkeko family until she was 10 months old, when she was caught and put in a pen with another young male takahē. Within a day, the pair had struck up a relationship, with Mana showing no signs of pūkeko imprinting, although Judah observed that she seemed a little more wary than an average takahē. Pūkeko and takahē behave differently with their chicks when they respond to danger. Pūkeko are warier and leave their chicks, which know to hunker down and be quiet. At least one takahē parent, on the other hand, will stay with a takahē chick to keep it safe. Mana's response to being abandoned by her foster pūkeko parents was to call loudly in distress, which was possibly not a great strategy. She died aged two and a half, without producing any chicks.

Another interesting pūkeko relationship involved a yearling takahē called Robin, also one of Squeak and Taku's offspring. He had been moved from Kapiti to Mana Island but found himself without a partner after the female he was meant to pair up with died following a fight. Instead, he attached himself to a pūkeko family, staying with them for eight months until he was moved to Maud Island. During that time he helped feed the pūkeko chicks, and took on a few pūkeko habits: his call was a high-pitched squeak rather than a low 'oomph'; he adopted the pūkeko alert posture, with outstretched neck and rapid tail-flicking; and he even took to roosting in low mānuka bushes rather than on the ground. However, on Maud Island with a takahē partner Robin soon dropped these new habits and went back to being a takahē. He lived for 14 years and fathered five chicks.

Another interesting experiment that Judah carried out with takahē and pūkeko on Mana Island was to rate their response to a predator. He assumed that pūkeko would be more responsive to a threat, as they were relatively recent arrivals in Aotearoa and had proved much more able to live in the presence of predators such as stoats than had takahē, which had evolved in the absence of mammalian predators. The experiment involved a radio-controlled car, a taxidermied stoat and a similar-sized cardboard box, with either the stoat (the predator) or the box (the control) mounted on the car. The set-up was initially hidden from view in a large box, and food pellets were scattered in front of it to lure the birds in close. Then the stoat or the cardboard box would be driven in and out of concealment for a couple of minutes to see how the birds reacted. Both species responded by calling, tail-flicking and increasing their vigilance, but unsurprisingly, pūkeko were more alarmed: they called more, flicked their tails a lot and were very vigilant, especially to the stuffed stoat. Mana, the takahē reared by pūkeko, was more responsive than another young takahē that had a normal takahē upbringing, and Judah speculated that it might be possible to use cross-fostering to pūkeko to raise takahē that are more streetwise when it comes to predators. It's a novel idea that has never been taken further, partly because it's very time-consuming and partly because the island takahē already had homes that were free of predators.

Perhaps the most significant fact to come out of island research at this time was the realisation that takahē on islands were not producing as many chicks as their Fiordland counterparts. There were high levels of infertile eggs and only a small number of chicks survived, a situation that was getting worse over time. This was despite the islands being free of predators and having a more benign climate, which had originally led researchers to speculate that island takahē would produce an abundance of chicks. The news wasn't all bad – if a chick made it to four weeks on an island, it had a much better chance of

Mana Island rangers Nick Fisentzidis and Genevieve Spargo vaccinate Tureiti and give her a health check before she is transferred to the Murchison Mountains, via Burwood.
DALLAS BISHOP

surviving to breed than a chick in the Murchisons, and island adult takahē also had a better chance of staying alive.

Curiously, the figures for egg infertility and low chick survival were worst for first clutches of eggs, and second and even third nesting attempts were more successful. However, the researchers couldn't come up with a convincing explanation for this.

There was lots of speculation about possible reasons for the disappointing island productivity – were the islands too dry and the eggs unable to regulate how fast they lost water during incubation? Or was the long, wet grass making the small, downy chicks cold and wet? Was the issue to do with how old the birds were, or with them being moved? Was it to do with the quality of food, such as a lack of a vital nutrient in the grass? Were pesticide residues still present from the farming days?

In the end, the best explanation seemed to be genetic. Small island populations formed from a small number of founder birds and with plenty of opportunities for close relatives to breed with one another had resulted in inbreeding. With her 62 grandchicks, super-breeder Squeak had created quite a genetic monopoly on the islands – and that is never a good thing.

CHAPTER TWENTY

TIRITIRI MATANGI ISLAND'S AVIAN AMBASSADORS

The sun is shining and the sea is calm as the ferry heads from Gulf Harbour on Whangaparaoa Peninsula for the last leg of its journey to nearby Tiritiri Matangi. There's an excited hum from the crowd of day trippers on board as the ferry slows and draws alongside the jetty. We file ashore past the welcoming DOC rangers and disperse along different paths for a day of walking, exploring and experiencing nature. There are secluded and shady bush trails, feeders bustling with hihi and bellbirds, reclusive kōkako singing from the treetops, and vistas of the Hauraki Gulf and the historic lighthouse at the island's southern end to admire. Like most of the visitors, I eventually find myself near the small shop, run by volunteers from the Supporters of Tiritiri Matangi, where people are deciding on souvenirs or eating their lunches, seated at picnic tables or sprawled on the expanse of grass out front.

A wave of excitement breaks the quiet hum of conversation, heads turn, and people clutching cameras and mobile phones rise to their feet. Island royalty has arrived. Unfazed by the attention, Greg the takahē makes his stately way among the legs, ignoring the squeals of delight from kids and murmurs of appreciation from the crowd. Greg is one of Tiritiri, and possibly New Zealand's, best-known birds and he's the island's biggest drawcard. This is the first time many of the visitors have seen a takahē and he doesn't disappoint – he is striking, imposingly large, fearless. No doubt he's hoping for a surreptitious handout, but this crowd has been well briefed not to feed the locals. Greg soon tires of the fuss and heads across to the grass to resume feeding.

I watch as his slow, deliberate wanderings and pluckings bring him close to where I'm sitting. He boldly walks right up to me and I stretch out an empty hand as a gesture of welcome. Greg investigates it and is clearly inviting me to scratch his head. I'm more than

The beach on Tiritiri Matangi was one of Greg's favourite hangouts. He was even known to take a swim in the sea. **STEVE TODD**

happy to oblige. It's delightful to be permitted such intimacy with a bird as rare as this, and completely on his own terms. After several minutes, his itch has been alleviated and he strolls off. By the time Greg disappears about half an hour later, he has bestowed his attention on everyone present, and we all feel richer for having shared our island experience with one of New Zealand's most charming birds.

Greg isn't the only takahē I encounter on my trip to Tiritiri Matangi in 2009. While none of the others are habituated to human company and quietly disappear if people get too close, I appreciate them just as much for their wildness and for treating us human visitors simply as part of the island's living backdrop.

****

In his 18 years on Tiritiri Matangi, Greg the takahē became a popular poster boy for the species. He was relaxed and gregarious with people, and took it upon himself to meet and greet day visitors to the open sanctuary in the Hauraki Gulf / Tīkapa Moana, sometimes even wandering down to the jetty to form part of the welcome party. As a result of his nosy nature, his footprints are unintentionally immortalised in some concrete near the bunkhouse. I know I was just one of thousands of people who had the privilege of a close encounter with him, and while he is long dead, his memory remains strong.

Greg was much more than just a social butterfly. He was an island resident from 1994 until his death in August 2012, and in that time he had three partners and sired seven offspring. Greg was the son of two unknown wild Murchison Mountains takahē, and he hatched in the Burwood rearing unit on 12 December 1992. He was fed with a puppet, fostered out with other chicks and remained at Burwood for the winter. The next summer there were rumours that the one-year-old had taken up with a mature female in the top pen, but he was soon returned to his rightful spot with a group of other yearlings.

In May 1994, Greg was relocated to Tiritiri Matangi, where he paired up with fellow newcomer Pounamu. She was the same age and had been raised by her parents, Yellow and Green, at Burwood. Young and inexperienced as they were, the eager pair hatched and raised two chicks that summer, Maia and Whakama. Their son Kristin followed the next year, with daughter Ahikaea in 1997 and son Bellamy in 1998.

In 1999, Greg was on his own as Pounamu had left him for their son Whakama. Shacking up with a close relative can be hard to avoid in a small population where there is simply not much choice. This helps explain why brother and sister Kristin and Ahikaea also found themselves together. Until that point, the relatively new island populations had been left to their own matchmaking, with little or no attention being paid to the genetic consequences. From the early 2000s, however, there was a change in focus to move birds around, to minimise further inbreeding, to top up the Murchison Mountains population, and to establish new populations at different sites, which explains where Greg's next offspring ended up. Post-Pounamu, Greg was briefly with Maudie, then back with Pounamu, before hooking up @Dot. She was a born and bred Tiri girl, which may explain why their 2003 daughter was called Tiri – although she was destined for release in the Murchison Mountains. Greg's next partner was Cheesecake, another Burwood female, and after producing an unnamed chick of unknown sex in 2008, they had a daughter, Ella, in 2009, Greg's eighth and final offspring. Ella was relocated to nearby Motutapu Island in the early days of establishing a population there. In between the arrivals of his two youngest chicks and following his retirement, Greg had plenty of free time in which to perfect his socialising skills with humans.

Greg's family is another good example of how not every takahē contributes equally to future generations. Only three of Greg's eight offspring went on to have chicks, and half of

Rank pasture grasses on Tiritiri Matangi Island suit the resident takahē. MARTIN SANDERS

those were from sibling matings. Sister and brother pair Ahikaea and Kristin produced four together: Blackwatch, Sapphire, Kristina and Mary. Ahikaea then paired up with another brother, Bellamy, and they had three offspring. Finally, Ahikaea moved on to new partner Montague, who was not a close relative, and had a further seven chicks. Montague is a familiar name from the Squeak and Taku dynasty and had previously fathered a daughter, Ihi, with his mother (see Chapter 14). Ahikaea's grand total over a 13-year breeding career was 14 chicks. Greg's eight remaining grandchicks were fathered by son Bellamy with mates Mingi and Paku (see Chapter 31).

That pattern of just a few birds from each generation breeding successfully is pretty universal. Only five of Greg's 22 grandchicks are known to have bred, and he's now great-grandfather to their 31 offspring. At this point, actual numbers become hazy, as the wider Greg whānau is now scattered across most of the takahē sites – including growing numbers that have been returned to the wild population in the Murchison Mountains, where their breeding is not monitored. But according to the pedigree record at least, Greg's family tree now spans five generations, and in the managed population he has 26 great-great-grandchicks so far. He may be long gone, but the future of Greg's genetic dynasty seems very secure.

CHAPTER TWENTY-ONE

MASTS IN THE MURCHISONS

There were two local mast-seeding events in the Murchisons at the turn of the century, in successive years – 1999 and 2000 – which is unusual. Back-to-back mast events are rare because plants generally need a year to recoup resources and energy. The first mast was a double whammy in another sense, too, as the snow tussocks were masting as well as the beech. Botanist Bill Lee had been monitoring snow tussock flowering in the Murchisons since the mid-1970s, and in late 1998 he observed that more than 80 per cent of a range of snow tussock species were in flower – the highest proportion he had ever seen. Bill also noted that the warmest summer on record might trigger another mass tussock seeding the following summer. At the same time, seedfall monitoring showed a big seed year for all the local southern beech species, with up to 12,000 seeds being recorded for each square metre of forest floor – it wasn't the biggest year on record, but it was significant.

The flowering of this first beech and tussock mast began in spring 1998 and was part of a widespread mega-mast event. Seed ripened and began to fall in autumn 1999, and rodent numbers started to climb as winter set in. There were a lot of takahē deaths that year, but only one out of 12 was attributed to predation.

Anticipating an increase in stoat numbers from spring 1999, the takahē team established a small network of stoat traps at four sites in the eastern Murchisons: Mystery Burn, Point Burn, Takahē Valley and the Te Ana-au Glow-Worm Caves. In just a few months, the 95 traps caught 144 stoats.

As all this was unfolding, the beech forest in the Murchisons was experiencing a second, smaller mass-flowering event. Only mountain beech flowered this time, dropping 5,423 seeds per square metre in Takahē Valley, but it was enough and by spring 1999 stoat numbers were climbing again. The 1999/2000 mast was not such a significant event in terms of stoats – only 60 stoats were trapped, and just four birds were found dead, two probably as a result of stoat predation. In an average year, about 3 per cent of takahē eggs disappear

Mountain beech in flower in a mast year. **ROD MORRIS**

due to stoat predation, and in these two mast years that number climbed to 6 per cent. But despite the two back-to-back stoat plagues, the Murchison Mountains birds escaped pretty much unscathed and the takahē population remained steady, with 216 birds in 1999, and 221 in both 2000 and 2001.

However, the two stoat plagues and associated takahē deaths did give the takahē management team – and biodiversity rangers in the DOC office in Te Anau who worked with other rare species such as kiwi, rock wrens and mohua in the same area – pause for thought. Money was found to create a low-intensity, landscape-scale trapping network across the southeast corner of the Murchisons, an area of 15,000 hectares that covered about one-third of the Special Takahē Area and included the valleys that were already being trapped. It was a good example of research and management working hand in hand – a key part of the takahē population would gain protection, while an untrapped area to the north remained free of traps for comparison.

Work began in 2002 to mark out a 120-kilometre grid of routes that mainly followed valleys or ridgelines. The grid covered both forest and alpine areas, as stoats have been caught from altitudes of 200 metres above sea level on the shore of Lake Te Anau up to 1,480 metres, and rats have been trapped as high as 1,440 metres above sea level. In addition,

Mast seeding of snow tussock in the alpine and subalpine zone is not always synchronised with nearby beech masts. JAMIE HICKLING

a recently completed radio-tracking study of stoats at several Murchison Mountains sites had shown that those animals living in forested valleys didn't venture up into the alpine area and vice versa, so it wouldn't be sufficient to trap solely in the forest and hope that stoats wandering down out of the alpine zone would be caught. In total, 720 trap boxes were placed at 200-metre intervals along each marked route, achieving a density of one per 21 hectares. Each box contained two Mark 4 Fenn traps, instant kill traps capable of killing both rats and stoats. The traps were baited with a fresh hen's egg and rabbit meat, which is a well-proven recipe for attracting stoats – the egg is a visual lure, while the smell of rabbit meat is irresistible to a hungry mustelid. Starting in winter 2002, the traps were checked and rebaited four times a year, each August, November, February and May.

The traps did their job – over the next four years, they caught 590 stoats and 490 rats. Stoat numbers peaked significantly each summer, with 81 killed in February 2003, for example, and dropped in winter, when as few as 10 were caught. Many of those trapped in summer were young, inexperienced stoats. Interestingly, as stoat numbers trended slightly down across those years, rat numbers climbed fivefold, the inference being that fewer stoats meant that fewer rats were being eaten. Despite several moderate mast-seeding years, there were no subsequent stoat outbreaks and reports from the time noted that it was still too soon to tell if the stoat trapping was proving protective for the takahē.

The takahē population trajectory continued a pleasing upward climb in the early 2000s. The number of island birds swelled. Rangers monitored the Murchison Mountains takahē each spring and summer, and collected 'excess' eggs and chicks for captive rearing at the Burwood unit, which were duly returned back to the wild the following spring. By mid-2005, there were 168 takahē in the Murchison Mountains, out of a total takahē population of 287. This was an increase of more than 20 per cent in the overall population since the turn of the century. Everyone was feeling optimistic.

A network of hundreds of stoat traps now covers much of the Murchison Mountains.
DEIDRE VERCOE

CHAPTER TWENTY-TWO

THE 2007 STOAT PLAGUE

The summer of 2004/05 was a hot one and much warmer than the previous year. As a result, tussock watcher Bill Lee predicted a 'huge flowering' of *Chionochloa* the following summer.[1] Sure enough, in spring 2005 a combined tussock and beech mast began its slow, lethal unfolding. Rather disarmingly, it wasn't a particularly massive mast year. Seventy per cent of the snow tussocks in Takahē Valley seeded; this was high, but less than in the 1998/99 mast, which had had little impact on takahē. By late autumn 2006, the beech seedfall monitoring in Takahē Valley showed 4,116 mountain beech seeds per square metre of forest floor – the fourth-highest tally in 25 years of monitoring, but nowhere near numbers for 1979, when more than 7,000 seeds per square metre were recorded. As the eight seedfall trays in Takahē Valley measure only mountain beech, the takahē team also monitor beech seedfall at a couple of other sites in Fiordland. Seedfall trays in three locations in the Eglinton Valley, for example, reflect what red beech and silver beech are up to. In autumn 2006, the counts came in at more than 6,000 for red beech in one valley and nearly 4,000 for silver beech in another.

The rodents began responding to the huge pulse of food – mice numbers were on the rise and, although lagging slightly behind, rats were on the increase as well. There are no records for mouse numbers, but catch totals from the 15,000-hectare trapping network in the eastern Murchison Mountains reflect what was happening with the rat and stoat populations. In summer, as seeds were ripening but weren't yet edible, 34 rats and 44 stoats were trapped. Stoat numbers remained the same in autumn, but 75 rats were caught as the population began to respond to seedfall and the increase in invertebrate and mice numbers. Over the winter of 2006, the rat population continued its upward trajectory, reaching 145 catches, while the stoat catch actually dropped to just 12. The Murchison Mountains takahē population, meanwhile, remained much the same as it was a year earlier, at 165 birds.

By spring 2006, a year after the mast began, the rat catch reached 319, while stoat

Flowering mid-ribbed snow tussock. ROWAN HINDMARSH-WALLS

numbers remained steady at 42. Even though the beech seed had disappeared, there were clearly enough mice and invertebrates still available to bolster the continuing explosive growth of the rat population, and the summer tally of 397 catches meant that more than half of the traps in the network had caught a rat. At this stage, female stoats had all given birth, juvenile stoats were on the move and there was no shortage of rats for them to dine on, a situation reflected in an increased stoat catch of 67.

By autumn 2007, the rat population had passed its peak. A lower rat tally of 210 reflected the fact that high numbers of rats were now struggling to find food as mice and invertebrate populations had already crashed. Conversely, with plenty of rats still available, these were good times for stoats, which clocked in at 166 catches. Their numbers continued to climb, with 205 stoats caught in traps over winter. However, with just 93 rat catches over the same period it was clear that these were increasingly lean times and there was every expectation that stoat numbers would crash, ending the effects of the mast.

At 168, the takahē population had stayed steady for the third year in a row and it seemed the birds' size had once again kept them out of harm's way. But this mast year had a nasty time lag built into it, and its lethal effects were yet to be seen.

In early spring 2007, two years after the beech and tussock first flowered, takahē rangers

Stoats will destroy eggs, and are capable of killing adult takahē. ROSS CURTIS

headed into the Murchison Mountains to monitor takahē breeding. On 8 September, they found the bodies of Darwin and an unidentified takahē. Four days later, they found Heki and Taurangi, also dead. All four birds were adults, and all showed signs of stoat predation. The carnage had begun.

In October, rangers found the bodies of Matt, Miller, Rakimanu and Pilindi. In November, they discovered Kopere and two unbanded birds. Arwin and Gepetto were found dead in December. In January, it was Kama. The bodies of Joe, Stuart, Attenborough and Waimanu were retrieved in March 2008.

All up, the rangers found 23 dead takahē in a six-month period, and 18 of the deaths were attributed to predation by stoats. The dead birds were not chicks or juveniles – all were adults in the prime of life. There were males and females, some hand-reared at Burwood and some raised by wild parents in the Murchison Mountains.

And that was only part of the story, as many bodies weren't found. The known adult takahē population in the Murchisons dropped by 43 per cent, from 168 birds to just 93. This was the lowest number on record – less even than the historic low of 1981, when management first began. And it could have been worse – much worse. There was a profound difference in the survival rates of takahē living in and out of the trapped area. In the untrapped area, the takahē population dropped by 58 per cent, whereas the decline in the trapped area was just 9 per cent. Any last lingering doubts about the impact that stoats could have on takahē were finally dispelled.

No one knows exactly why this mast season ran for so long and turned out to be so deadly. It was probably the combination of a few things. A large tussock mast alongside a beech mast fuelled a very large, long-lasting mouse population explosion that then fed an enduring rat plague. The sheer numbers of rats overwhelmed the trapping network, and a trap filled with a dead rat is no longer available to catch a stoat. And takahē managers believe that, even as their numbers were in freefall, the desperate stoats switched prey when they ran out of rats to eat – and the prey they fixated on were takahē.

The slaughter of so many takahē by stoats and the significantly better survival of birds in areas that were being trapped led to a prompt decision by DOC to immediately increase trapping effort. The area to be trapped would rise from 15,000 hectares to 50,000 hectares, enough to help protect almost every takahē in Fiordland. As before, trap boxes would be placed at 200-metre intervals along valleys and ridgelines, the only difference being the use of newer DOC150 traps instead of Fenn traps. There were now 1,600 traps that needed to be checked four times a year.

Predation has wide-ranging and sometimes long-running impacts. The year after the devastating stoat plague, the Murchison Mountains takahē population was in a state of social disarray. There were fewer breeding pairs than previously, as many birds had lost a mate and were in the process of trying to find a new one. With just 94 birds in the population – only one more than the year before – spread across 50,000 hectares, the odds of bumping into a single bird of the opposite sex had dropped considerably.

During the stoat plague that began in 2007 and ran well into 2008, there were 74 takahē deaths. ANDREW DIGBY

Single takahē started walking and, failing to meet a suitable bird, simply kept on walking. In an average year, there are usually a couple of reports from trampers and hunters who claim to have seen a takahē outside the Murchisons. In 2008/09 there were 12 reports of different individuals, some well away from the Murchisons. The best solution to anchor takahē in the comparative safety of the Murchison Mountains was to have a full house, with most available territories occupied, thus ensuring birds could find one another. The takahē team thought that 190–200 birds would fulfil this aim, without creating too much competition for space and food.

In this regard, a significant change in management practice had already been trialled on a small scale during the 2007/08 season. Since takahē populations were first established on various islands, it had been policy for that traffic to be one-way – a Fiordland bird could go to an island, but an island bird couldn't be sent in the opposite direction. It was a way of managing unknown disease risks. In addition, the idea was that the birds would acclimatise to the warmer, milder island environments, and as a result become unsuited to the much harsher Fiordland climate. But this season, for the first time, three island birds joined Burwood birds being released to boost the wild Murchison Mountains population. The three birds had first spent the winter at Burwood, learning how to eat tussock and *Hypolepis*, and getting used to the cold and snow. Once released, two at least survived the stoats, while the third went missing after losing its radio transmitter. This was enough to embolden the takahē team to continue with this approach, especially now that the Murchison Mountains takahē population needed all the help it could get.

There was also a small ray of hope from the Murchison Mountains: in the summer of the stoat plague, more than 90 per cent of the takahē eggs that were checked were fertile – the highest proportion on record. And the weather had been mild, which contributed to a higher-than-average hatching and survival rate for those chicks that escaped the attention of the hungry mustelids.

When it came to the 2007/08 breeding season, the takahē team were disheartened by yet another periodic plummet in the ongoing rollercoaster of bird management, which saw the overall takahē population drop by a quarter. But the rangers were relieved that they did have a growing takahē insurance population on the islands, where numbers had climbed to 89. With the newly expanded trapping network underway, the plan was to carry on hand-rearing takahē at Burwood and release those juveniles into the Murchison Mountains along with some island birds, to rebuild the population to capacity.

Overall, the 2000s were a torrid time for the takahē team. Ten years of hard work saw the overall takahē population end up almost where it had started – at 230 birds. There were just three years of reasonable growth. Everyone hoped the next decade would not be so dire.

A richness of rails

A rich variety of flightless rails was a distinctive feature of the pre-human New Zealand avian fauna, and it was one it shared with numerous islands across the Pacific.

Scientists have enthused that one of the largest radiations of island birds is that of rails. Today, there are about 150 or so species of rail worldwide, and of these about 30 are flightless. However, evidence from fossil bones shows that there were once hundreds – and perhaps as many as 2,000 – flightless rail species spread across the Pacific. Even the smallest island was home to between one and four species of endemic flightless rails. As American avian fossil expert David Steadman put it, 'it was one of the most spectacular examples of avian evolution'.[2]

All of these unique species disappeared in the wave of extinction that accompanied the spread of humans through Melanesia, Micronesia and across Polynesia, beginning in the western Pacific about 50,000 years ago and finally completed with the settlement of New Zealand about 750 years ago. The rails were an easy meal – often large, unable to flee fast, unused to mammalian predators, simple to hunt and, above all, tasty. As Steadman also pointed out, the group now had another spectacular claim to fame: 'at the species level this is the largest single extinction event ever detected for vertebrates'.[3] Yet, because no one was recording the disappearances as they happened, and because few people have gone looking for the evidence, the great Pacific rail extinction is little known.

Not your usual radiation

When it comes to island radiations, even biologists don't tend to think of rails. Instead, other groups of birds are considered to be the classic examples. Darwin's finches in the Galápagos Islands are one such group. Some 2–3 million years ago, a single ancestral finch arrived on the newly formed islands, from either South America or the Caribbean. Over time, this single species evolved, or radiated, into the 15 Galápagos finches know today.

A similar thing happened with moa, with nine large flightless species evolving from a single small, flying species that reached New Zealand. On the ratite genetic family tree, the closest living relative of the moa is the small flying tinamou from South America, so it seems likely the moa ancestor resembled that species.

In both the Galápagos finches and moa examples, what you have is a classic well-branched family tree with a single ancestral trunk. Where the story of rails on New Zealand and other Pacific islands differs is that they didn't all evolve from a single ancestor arriving at one point in time. Scientists have instead speculated that there were repeated arrivals of the same ancestral species over time, with each subsequent arrival soon evolving into a new flightless species. The ancestor would have been a supertramp (see page 110) such as the banded rail, which has a great ability to disperse and reach even remote island groups, and is predisposed to give up flight if conditions are right.

A variation on this theory that the latest genetic evidence seems to support is that there is a sequential linear evolution from the original species to a new one, which in turn rapidly evolves to another new species, and so on. On this family tree, the original trunk grows one branch, which in turn sprouts a single branch. Under this scenario, the founding supertramp may have arrived only once or perhaps twice, rather than multiple times.

Either way, the small islands of Oceania have been dubbed 'nurseries of biodiversity' by researchers for the way they spawn new species. And on islands, many of those new species are flightless giants.

Island life

The evolution of flightlessness

There are a number of related reasons why island birds – and especially rails – often become flightless. Partly, it is down to the bird itself, and partly it is down to the island environment the bird finds itself in.

Continents are large and are home to a wide variety of species that fill the available niches. There is a rich diversity of plants and invertebrates, which provide plentiful food. But there are also many hunters, especially large mammalian ones, which hunt on foot using smell to track down their prey. As a result, it is an advantage to be able to fly to stay out of the way. A flying bird needs to be light so that it uses as little energy to fly as possible, and light often means small. It supports this mobile lifestyle with a diet of high-energy food such as fruits and insects to power its large flight muscles.

Islands, on the other hand, are often small. They are also isolated and remote, and only species that are either capable of reaching them under their own power, or that have the fortitude or luck to survive long accidental journeys, make it to their shores. Plants with light wind-blown seed, birds, invertebrates and some reptiles are better able to survive being blown or rafted across vast expanses of ocean than most mammals, except for bats.

Because they are difficult to reach, islands end up being home to only a small number of species, which occupy only some of the available niches. On the plus side, this leaves plenty of opportunities for any newcomers to exploit.

The lack of mammalian predators is crucial – in their absence, birds can save energy by giving up flying and adapting to a life on the forest floor. Down here, they can also effectively hide from any birds of prey, which hunt using sight and sound. Studies have shown that on

Two female takahē fighting over a male on Tiritiri Matangi Island. MARTIN SANDERS

islands even birds that continue to fly have smaller flight muscles but longer legs. And the fewer predators and the smaller the island (and those two factors are connected), the more likely birds are to become flightless. More than a thousand lineages of birds have become flightless on islands, from parrots and ducks to owls and rails. The rails in particular seem to lose flight readily.

Of course, another part of the equation is a bird's diet. A bird that hovers to get nectar from flowers or hawks for flying insects on the wing will need to keep flying to find food. But for herbivorous birds such as New Zealand's takahē and kākāpō, there was plenty of food on the ground that they didn't need wings to reach.

Gigantism

Increasing size often goes hand in hand with flightlessness for island birds – for example, consider the moa, kākāpō and takahē in New Zealand. These all evolved to occupy the grazing herbivore niche that animals such as goats and deer fill on continents, eating leaves, grasses and some fruits. Walking is the most efficient way to reach this food, which is low in energy and difficult to digest, requiring a larger – and hence heavier – gut to deal with it. For the moa, kākāpō and takahē, flightlessness and gigantism were the inevitable energy-saving consequences of their new diet.

Becoming a giant has other benefits. For one, it means a species is better adapted to cope with colder environments – such as those experienced during the frequent glaciations that have been a feature of the past 2 million years. At the same time, the flightless giants evolved to live longer, lay large eggs and breed slowly. And in the absence of predatory mammals, island birds became much more naive and less alert to the possibility of being hunted. This has made them very vulnerable when predators are brought to their island habitats, most often by people seeking to settle there themselves.

PART 4

ADVENTURE

2010s and 2020s

Kai mau te takahē mō ake tonu atu.

The takahē will be around forever.

– Michael Skerrett (Ngāi Tahu)

CHAPTER TWENTY-THREE

CAPITAL CITY TAKAHĒ

'Wow, that's awesome.'

'It's so big.'

Unperturbed by the small, admiring crowd and the clicks of their cameras, Nio wanders out from under a bush and approaches the feeding hopper. She lowers her head and with a deft flick opens the lid. It gently rests on the top of her head as she carefully picks out a pellet.

'Look! Another one.'

An excited child has spotted Orbell, strolling out from under the same bush to join his mate at the feeder. He grunts quietly, companionably pushes his head in next to hers and for a few minutes they feed together.

The crowd grows as more people turn up.

Snack time over, the pair begin to graze the long grass. They move slowly, a couple of metres apart, their route taking them closer to the watchers, who are separated by a low rope barrier.

It is spring 2017, and the ageing takahē pair has just passed the two-month anniversary of their move to Zealandia Te Māra a Tāne, a fenced sanctuary in central Wellington that is a mere 10-minute drive from Parliament and the Beehive.

I sit down at a picnic table nearby to watch and wait. As the crowd thins, Nio and Orbell wander under the rope and across the path towards me. I can hear their gentle, rhythmic 'oompf' calls as they stay in touch with each other; it's a deep sound resonating from their chests. They feed steadily, snipping the longer blades of juicy pasture grass with sharp-edged bills that work like secateurs. They drift closer, allowing me to admire their shaggy feathers, no good for flying but a perfect thermal blanket for cold weather. Seen in sunny close-up, the colours are vibrant and iridescent. Their belly and flanks are deep purple hues, with an upper blended border of almost turquoise-blue, and the wide band of khaki-green and brown feathers drapes across their back like a natural kākahu cloak.

Orbell trying to mate with his partner, Nio, at Zealandia. JANICE MCKENNA

Nio and Orbell are advocacy takahē, spending their twilight years being ambassadors for their species. Once they had had enough offspring and were considered to have made an adequate genetic contribution to the species, they were retired to the mainland to free up their Mana Island territory for younger breeders. In the process, they will allow many more people to have their first sighting of one of New Zealand's most endangered birds. In 2017 Nio is 13 years old and her mate Orbell is 17 years old. This is a second 'arranged marriage' for each bird, and despite being older parents the partnership has been more fruitful for each of them than their previous relationships. In 2013, they produced two sons, Brodie and Bon. Then in 2015 they had a daughter, Tara, and in 2016 another daughter, Tupuānuku.

Previously, Orbell – named, of course, after Geoffrey Orbell – had been paired with Bunchy, who was four years older than him. They produced two sons: Grant in 2002 and Geoffrey in 2007. Nio's previous mate was Port, and the only two chicks they produced died around hatching in 2012. After Nio, Port went on to a much more successful pairing with Astelia, producing four chicks together.

T2 and Puffin were Zealandia's first pair of retired takahē. CHRIS HELLLIWELL

Despite being used to a very quiet existence on isolated Mana Island, Nio and Orbell have taken to their new life in the spotlight of public gaze at Zealandia like ducks to water. They inhabit an area known as the takahē lawn. Zealandia sits in a large valley that was once the main water supply for the small town of Wellington and has two deep reservoirs. The takahē home is bounded on one side by the lower reservoir, and it backs onto a wetland on the other side. There are plenty of bushes and tall grasses where the birds can hide from sight, and an open lawn area in the middle that is open to the public.

After the move here, Nio and Orbell soon realised that rangers and Zealandia volunteers make daily visits with food. They learnt to rush to the gate when they heard the regular footsteps approaching on the floating pontoon that marks the northern entrance to the grassy area. The takahē have free access to the lake if they desire it, and one morning the feed-out ranger was startled to hear a lot of splashing. It turned out that one of the birds was so eager to get to the food, it had taken to the water and was swimming towards him.

Nio and Orbell were retired from Mana Island to replace Zealandia's previous advocacy takahē, T2 and Puffin. This pair had also spent most of their lives on Mana. T2 got his

OPPOSITE: Puffin eating roots and stems of wetland plants. CHRIS HELLIWELL

unusual name because his father, Snow, who was an aggressive bird when it came to people, was nicknamed the Terminator. In terms of personality, however, T2 was nothing like him.

T2 hatched in 1994 and Puffin a year later. They weren't a particularly prolific couple. They produced three female chicks, although two of them – Bimbo and Te Mana o Kupe – survived only a couple of years and died without breeding. Matariki, however, who hatched in 2000, was moved to Maungatautari in the Waikato and went on to produce seven chicks there, before dying in 2016. The early 2000s were a boom time for takahē on Mana Island, with most pairs breeding well and the population reaching more than 40 birds.

Puffin and T2, however, were retired to Zealandia in 2011, after seven years without successful breeding. Although they would mate in spring, nest and lay eggs, they failed to produce any that were fertile. At Zealandia, the pair were a crowd favourite, and would happily wander right up to people sitting on the grass or on one of the benches.

In mid-2017, at the ripe old age of 21, Puffin became lethargic and was off her food. She was taken to The Nest Te Kōhanga at Wellington Zoo for veterinary treatment, where she was diagnosed with a gastrointestinal upset and possible a partially obstructed bowel. She seemed to be recovering well following medical treatment when she suddenly died.

The Takahē Recovery team soon decided to give Zealandia a new pair of birds – Nio and Orbell – along with a third to keep T2 company.

Twenty-year-old Ahikaea had recently lost her partner Montague at Auckland Zoo. She had quite a different history to T2. While he had had a long-standing monogamous relationship with Puffin and produced only three chicks, she had been a prolific breeder with three previous partners. She was the daughter of Greg, the well-known takahē from Tiritiri Matangi Island, where she also spent most of her life. She hatched early in the 1997 breeding season – hence her name, which means 'first of spring'. With her first mate, Kristin, she produced four chicks. With her second mate, Bellamy, she produced three more, then with Montague she produced a further seven chicks. However, being prolific didn't guarantee that the next generation would follow suit – only six of her offspring are known to have bred, and only two of those had more than one chick

While Nio and Orbell took over the takahē lawn area at Zealandia, T2 and Ahikaea were provided their own secluded area, away from the public gaze. They soon settled in with each other and enjoyed four months together before T2 died in April 2018 at the advanced age of 23. Ahikaea died two months later.

Older takahē pairs aren't expected to breed once they are moved to an advocacy site, but sometimes – to everyone's excitement – that is exactly what they do.

Nio and Orbell arrived in spring 2017 and it wasn't long before their minds turned to procreation. They courted discretely, mated in the evenings when all the visitors had left, and disappeared to make several trial nests. Eventually, out of sight, Nio laid and the pair began to take turns incubating. Only one bird was now seen, heading to the feeding hopper

Nio with her two-month old chick Te Āwhiorangi in 2019. CHRIS HELLIWELL

for pellets, browsing for a while on the lawn and then disappearing to feed closer to the nest.

The birds were mostly left on their own, although one quick check by staff confirmed a fertile egg. Just before Christmas, the pair's behaviour changed when the chick hatched. But there was cold, torrential rain over Christmas Day and Boxing Day, and soon afterwards the sanctuary staff noticed that the pair were no longer behaving as if they had a chick. No one knows for sure, but the small downy chick likely got wet and chilled. The sanctuary was delighted that their retired couple were still keen to breed so soon after moving to a new home, and the hope was that they would try again.

Nio and Orbell didn't disappoint. In 2018 they got off to an early start, and in mid-October the sanctuary announced excitedly that they were incubating an egg. The chick followed a month later, its arrival almost exactly coinciding with the rediscovery of the takahē 70 years earlier by Orbell's namesake. The takahē lawn and wetland area were closed off to visitors to allow the parents to raise their chick in peace and quiet, and the only people present were rangers delivering regular supplementary food parcels for the doting parents to feed the chick. The food was a special mash of moths, clovers and vegetables, developed after many years of takahē hand-rearing.

Visitors to the sanctuary were eager to see the new arrival. At a couple of viewpoints with small views of the takahē lawn, people peered through binoculars hoping for a glimpse of the family. And occasional photos released by Zealandia catapulted the chick to social media stardom. In the regular updates, it progressed from a tiny ball of black fluff with a distinctive white bill, to a larger ball of black fluff with just a white tip on its black bill. It graduated from baby mash to pellets, and from being fed everything by its parents to beginning to feed itself. And it became more confident about wandering away from its parents. At two months old, its feathers began to emerge through the down, bringing hints of green, blue and brown to its plumage.

When the chick turned three months old, it was caught for the first time by the rangers for a health check and vaccination. They took a blood sample so its sex could be determined, and gave it a unique combination of leg bands. On its left leg they attached a red plastic band above a numbered metal band (RM), and on its right leg an orange band above a black band (OB). The number on the metal band is the bird's identity for the rest of its life, while the coloured bands will enable it to be easily identified through binoculars from a distance. The takahē lawn was then opened to the public again, and a steady stream of people came to admire the chick, which was as nonchalant with the attention as its parents were.

By four months old a takahē chick is nearly the size of its parents. It is fully feathered, although a bit drabber than an adult takahē. Its bill is a pale red and its legs are a pale reddish orange; both of these darken and intensify in colour over the next year.

In mid-April 2019, the five-month-old chick – a male – was formally named Te Āwhiorangi by Wellington iwi collective Taranaki Whānui ki Te Upoko o Te Ika. The name means 'the encircler of heaven' and references a sacred pounamu adze, or cutting tool, that is said to have been used by the atua, or god, Tāne to cut the sinews that bound Ranginui, the sky father, and Papatūānuku, the earth mother.

Although his parents are advocacy birds, they were once a part of the takahē diaspora spread across many sanctuaries and islands. Te Āwhiorangi, however, will have a different destiny. The takahē team made the decision that, with his low genetic value (he has plenty of relatives, so his genes are already widespread), he would be an appropriate addition to the wild population in the Murchison Mountains. Before being moved to subalpine Fiordland, however, the young bird first had to acclimatise to the colder climate and learn how to eat tussock and *Hypolepis* fern roots. When he was six months old, he was sent from Zealandia to the bird bootcamp at the Burwood Takahē Centre, to spend his first winter receiving tussock training and generally learning how to be a real southern takahē with a pair of foster adults. That was followed by a stint as helper to a breeding pair. Then, well trained, 18-month-old Te Āwhiorangi was released to make his way in the wild takahē world.

Sponsorship

The first corporate sponsor of the Takahē Recovery Programme was the Flight Centre, in the late 1990s. The next was hardware chain Mitre 10, which became involved in 2005 after 10-year-old Southlander Sophie Smith wrote them a letter asking for their support. Mitre 10 Takahē Rescue ran until 2016, with the company providing money, building supplies and keen volunteers. For the following few years Mitre 10 continued to provide building supplies for projects such as new pens at the Burwood Takahē Centre.

Since 2016, civil engineering firm Fulton Hogan has been the Takahē Recovery Programme's national partner, helping it with funding and practical support such as constructing new pens at Burwood, and in return being invited to assist with events such as releases of birds into the wild. Individual staff appreciate the chance to help in practical ways. In 2018, for example, Fulton Hogan employee Richard Harrowven won a staff challenge to design a portable lightweight pen that could be used by takahē rangers to capture birds out in the field for health checks. Richard's design called on his engineering background and was a practical solution that DOC could use immediately.

Back at Zealandia, Nio and Orbell's thoughts turned to making chicks again when spring rolled around in October 2019, but even though they nested and laid an egg, nothing came of it. It was the same story in the 2020 breeding season. Then for a time in 2021 the pair were relocated to a different part of the sanctuary while the lower lake, which borders their wetland home, was drained to allow the eradication of invasive exotic perch. The birds settled well into their new digs and were often spotted by people walking outside the sanctuary, as their favoured feeding area was the lush grass growing along the fence line.

After the fish eradication was successfully completed, Nio and Orbell returned to wooing visitors down at the takahē lawn, but as this book goes to press it's likely they will be returning to their temporary territory for good. This will allow a younger breeding pair to take over the main territory, which should guarantee many crowd-pleasing fluffy chicks on display for years to come.

CHAPTER TWENTY-FOUR

A CHANGE OF DIRECTION

The start of a new decade in 2010 brought big changes to takahē conservation. The new direction was iterative, slowly building on all the previous research and management work, yet it was also a significant departure from some management practices that had been in place since the early 1980s. In a nutshell, the plan now was to do less but to do it in more places, to maximise the genetic health of the species and to let the birds themselves do more of the hard work. The intent was also to have more takahē in places where people could see them.

It was just over 60 years since the species had been rediscovered, and despite much research and 30 years of innovative and intensive management that had helped the species weather some significant crises, the takahē population wasn't increasing much. The overall population now stood at 230 birds, with just 109 of those in Murchison Mountains, where they were slowly rebuilding their numbers after the dreadful population crash of 2007 (see Chapter 22).

For staff of the Takahē Recovery Programme, the new decade meant dealing with the financial fallout of a major funding cut to DOC in 2009 that would lead, in 2011, to a drawn-out restructure and significant job losses in the department. By 2013, the takahē team in Te Anau found themselves partnered with the Kākāpō Recovery Programme, managed by Deidre Vercoe. Responsibility for managing deer and stoats in the Murchisons was passed to other units within DOC.

In the field, the biggest immediate management change was at the Burwood Takahē Centre, where the chick-rearing team warmed up the egg incubators and donned the famous blue hand puppets to feed little takahē chicks for the last time during the 2010/11 breeding season. None of the 14 chicks that hatched came from eggs collected from the wild population in the Murchison Mountains, as had been the practice for so long – instead, five eggs were from Mana Island and nine were from Burwood's six resident breeding

There is a range of secure pens at Burwood Takahē Centre, for breeding and training birds.
BENHI DIXON / DOC

pairs. And for the first time in 20 years, none of that year's offspring were sent to boost the number of Murchison Mountains birds, which were being left to their own devices again.

Since its inception in the early 1980s, the pioneering captive-rearing programme had produced hundreds of young takahē. At first, in the mid-1980s, these were used to populate island sites, then in the experiment to create a new Fiordland population in the Stuart Mountains in the early 1990s, and more recently, they were used to bolster numbers in the Murchison Mountains. But there had always been the question of whether this strategy was working as well as intended, and a review of how takahē were being managed brought the issue to a head. Several analyses proved something that had been suspected for quite a while: captive-reared takahē survived well in the wild in Fiordland, where they made just as many nests and laid as many fertile eggs as their wild counterparts that had been raised by their parents, but they were far less productive. When it came to hatching success and how many of those chicks made it to fledging, wild birds performed twice as well as captive-reared birds. Pairs formed by two captive-reared birds did the least well, while pairs made up of a wild bird and a captive-reared partner were less productive than two wild birds.

Does this mean that the entire hand-rearing programme was a failure? Not at all. It had indeed succeeded in its aim of supporting the Murchison Mountains population, and it had been a groundbreaking achievement that inspired similar hand-rearing programmes for other threatened species such as kakī. But it was clear that there were other, simpler ways of 'making' takahē. It would be straightforward to repurpose Burwood to become the 'engine room' of the recovery programme, a safe place where an increased number of takahē pairs could get on with making as many new parent-raised takahē as possible.

Work began at Burwood on building more predator-proof pens to accommodate more breeding pairs, and the previous season's chicks – which were now yearlings – were retained at the breeding centre instead of being shipped off to the Murchisons. They would be the basis of the enlarged captive breeding programme, producing chicks that could be moved out to the islands and to a growing number of small sanctuary sites around the country. The plan was for Burwood to have 20 breeding pairs by 2013, although the youthful age of the expanded population meant it would be several years before successful breeding really got underway.

The Murchison Mountains takahē population entered a more hands-off management phase. There would be no releases of young birds, to see how well the existing population did with landscape-scale predator trapping and sustained deer control in place. The spring and summer censuses, which saw rangers scouring the valleys and peaks to monitor the survival and breeding success of each pair, were done away with. Nest manipulation – destroying dud eggs to encourage re-nesting, and moving fertile eggs between pairs and out to Burwood – was no more. The new plan was to monitor a large subset of the population from the air.

By the end of the 2010/11 summer, 55 takahē in the Murchisons had each been fitted with a radio transmitter, which was attached to the bird by a backpack harness, with cord loops that fitted around each wing. (In more recent years, the harness system has been changed to one called a flight harness, which is less likely to rub against the wearer's wings.) If the bird is moving. it emits a signal at one speed; if the bird dies and stops moving – or if the transmitter falls off – the signal changes to a different speed. Traditionally, radio-tracking has involved a person with an aerial and receiver standing somewhere with a good field of view or sitting in a helicopter, dialling up each bird's radio frequency one at a time, then noting where a signal was coming from. A new bit of kit revolutionised all that. Sky Ranger is an automated system that can quickly scan a large number of frequencies. A fixed-wing plane fitted with a Sky Ranger GPS recorder and an external aerial flies a grid pattern over the search area and logs the location and 'state' (alive or dead) of each of the birds carrying a transmitter.

Ranger climbing out of the Snag Burn during a Murchison Mountains takahē census.
PHIL MARSH

The takahē team use Sky Ranger four times a year to monitor the survival rates of a large sample of birds, and then from this they can extrapolate the overall size of the population. The transmittered birds are visited once a year in late summer, so any chicks can be counted and banded and the transmitters replaced as batteries ran out. As birds die, new ones are added to the study to keep the sample size large enough. And every three years, the team ground-truths the Sky Ranger numbers with a complete population census on the ground.

Another significant change was a strong focus on advocacy, on getting the takahē story out to more people. For decades, Pukaha / Mt Bruce and Punanga Manu o Te Anau / Te Anau Bird Sanctuary were the only places in mainland New Zealand where the general public could freely visit and see a pair of takahē. Since 2006, however, Sanctuary Mountain Maungatautari in the Waikato had been home to a pair of retired island birds: Hauhanga from Maud Island, and Matariki from Mana Island. In addition, Aucklanders could easily see takahē out on Tiritiri Matangi Island, and smaller numbers of visitors encountered the birds on Mana, Maud and Kapiti Islands – although such interactions were incidental to the islands' main purpose of holding satellite takahē populations.

Things changed in late 2010, when the number of advocacy display sites was expanded. Hamilton and Guy, a retired breeding pair from Burwood, were sent to Willowbank Wildlife Reserve in Christchurch as display birds (see box). Then in early 2011, the fenced Zealandia sanctuary in central Wellington became home to a retired pair from Mana Island, T2 and Puffin (see Chapter 23). And in 2012, Dunedin's Orokonui Ecosanctuary welcomed Quammen and Te Hoiere, from Maud Island (see Chapter 31).

A growing issue that needed to be addressed was that the existing islands were getting very full. The hunt was on for new sites to establish takahē populations – preferably large places that could hold plenty of birds. The team drew up a list of 27 possibilities spread across the country and began to assess their suitability. There were plenty of things to consider, such as suitable food, safety from predators (especially cats), water sources, the ability of the site to anchor birds and not have them wander off, shelter from bad weather and so on. Secretary Island in Fiordland's Doubtful Sound / Patea was already being trialled, with young male Blaze and 10-year-old female Arizona, both Burwood-raised birds, relocated there in April 2009. It was hoped the two would pair up, but by June Arizona had moved to the tussock tops, while Blaze remained on the western coastline near where he had been released. Six more takahē joined them in 2010 and 2011. One of the 1800s takahē specimens had been collected on Secretary Island, and with its range of ecosystems – including coastal grassland, subalpine tussock and forest – there were high expectations for it. The birds survived, but there is a big difference between just surviving and having enough resources to reproduce. In 2015, with no breeding taking place, the birds were taken back to Burwood. One pair was considered genetically valuable and the takahē team were keen to ensure they were encouraged to breed.

Burwood's hand-reared legacy

Between 1982 and 1989, 104 fertile eggs and four chicks were collected from the Murchison Mountains for hand-rearing at Burwood. Due to health and quarantine concerns, none of the birds that were successfully reared were returned to the Murchisons. Fifty-eight young birds were released into the Stuart Mountains between 1987 and 1992, while most of the rest were sent to islands, especially Mana and Maud. A handful of breeding birds were kept at Burwood.

In 1990, the takahē team decided to boost numbers in the wild population, and the first cohort of six hand-reared birds was released back into the Murchison Mountains.

Between 1991 and 2009, field rangers collected a further 163 fertile eggs and 61 young chicks for rearing at Burwood. These were supplemented by eggs laid by three to five breeding pairs of takahē that were housed at the rearing centre. The great majority of Burwood's output in this period – a total of 210 birds – was released back into the Murchisons, with a few birds heading to various islands.

A resident adult at Burwood with its chick. CRAIG MCKENZIE

In August 2011, newly pest-free Motutapu Island in Auckland's Hauraki Gulf / Tīkapa Moana became home to four young takahē. The former farm had plenty of grass and a plan for revegetation that had takahē requirements in mind. More importantly, it was large and the takahē team were optimistic it could accommodate plenty of birds.

This expansion of the number of sites where takahē were living was part of a strategy to improve the genetic status of the species, to retain as many different genes in the population as possible, and it's this focus on genes that has dominated takahē management since 2011.

Hamilton

From his birth in 1992 until his retirement to Willowbank in 2010 as an ambassador for the species, Hamilton was a star of the Burwood Takahē Centre and a favourite of ranger Glen Greaves. Partner to Hamilton from 1996 was his younger mate Guy – and yes, she was a female despite her name. Guy was hatched and raised by her parents at Burwood, and died just five months before Hamilton in 2012.

Hamilton started life in the Murchison Mountains, where he was selected as one of the surplus eggs to be airlifted to Burwood. He was one of 12 chicks in the hand-rearing class of 1992, along with another male that would go on to achieve takahē stardom – Greg (see Chapter 20).

While many of Hamilton's classmates were sent to the Stuart Mountains as part of an ongoing effort to start a new population there, he was kept on as one of the centre's resident breeding and foster birds.

Hamilton was a large, handsome bird with a big personality, says Glen. Hamilton and Guy had the middle takahē territory in the large 75-hectare enclosure, which was shared by three pairs of takahē. When a ranger turned up at feeding time to fill the hoppers, Hamilton would be there on the boundary of his territory to greet them. Glen says he would 'also usher you out at the end'.[1]

'As a big bird he could be quite intimidating for some of our smaller staff and volunteers,' says Glen. 'He never actually did anyone any harm – just gentle nips to the leg if you weren't submissive enough!'

Between 1997 and 2008, Guy and Hamilton raised 11 of their own chicks.

Each season they were also foster parents to many hand-reared chicks, showing them how to forage for different foods and how to behave like a takahē. Their own chicks were all released into the wild in Fiordland.

In 2010, the Burwood team decided to retire the pair and free up their breeding territory. They became the first advocacy pair at the W llowbank Wildlife Reserve in Christchurch, where they lived for two years until their deaths.

Hamilton. DOC

CHAPTER TWENTY-FIVE

GENETIC MANAGEMENT

Tumbles was a special takahē, for several reasons. His parents were wild Murchison Mountains birds, and he was collected as an egg from the Snag Burn in November 2004 and taken to Burwood for puppet rearing. In 2010, after failing to father any chicks with three different mates at Burwood, he was moved to the Te Anau Wildlife Park, where he was described as laid-back and fond of food. He was partnered there with Kawa, who was not needed in the breeding programme because she carried genes that were already over-represented in the population. The pair became star attractions at the park. They were also excellent foster parents and from 2012 onwards raised 13 chicks.

But takahē ranger Glen Greaves, who had known Tumbles all his life, had particular reason to be sad on New Year's Eve 2020 when the 16-year-old had to be euthanised because of an untreatable leg injury. For not only was Tumbles a great character and father, but he was also – at that time – the number one male in the takahē population from a genetic point of view. That importance was because of his wild provenance and lack of any known relatives in the managed takahē population. Glen was disappointed Tumbles had died without any progeny – if he had harboured any unique genes that weren't represented in any other bird, then they died with him.

Glen's role in the Takahē Recovery Programme has, for the past decade, focused on maximising the genetic diversity of the population. Concern about the genetics of the takahē population goes back to the early 1990s, when the late Professor Ian Jamieson and various students from the University of Otago began researching the conservation genetics of a number of rare New Zealand bird species. Ian had a background in pūkeko behaviour and ecology, and was interested in the closely related takahē – especially the island populations, which were growing fast but were inbred and experiencing lower egg fertility.

The subfossil bone record suggests that while takahē were quite widespread in the eastern South Island, they have never been particularly abundant there, probably because their

Tumbles with foster son Timata, who later became part of the new Kahurangi population.
ANJA KOHLER

preferred habitat – a patchwork of grassland, with shrubs, forest edge and wetland – has never been widespread. This is in contrast to the primarily forest-dwelling kākāpō, which were once one of the most common forest bird species in New Zealand. Archaeological evidence from Māori middens supports the rarity of takahē – they were eaten sometimes, but never in large numbers. And because takahē are slow breeders, it wouldn't have taken much hunting pressure to cause a population decline, meaning they were soon less available as food. By the mid-1800s the takahē population already numbered just a few hundred birds, mostly in Fiordland, and it dropped as low as 111 known wild birds in 1982, which is when intensive management began. Low numbers such as this are known as a population bottleneck, although the takahē bottleneck was not nearly as extreme as that of the Chatham Island black robin, which dropped to a single breeding pair, or the kākāpō, which got to just 51 birds.

Populations that go through population bottlenecks are prone to inbreeding, when related birds start breeding with one another more often. This often leads to a loss of genetic diversity in the population as common genes become more common and rare genes are more likely to disappear. Genetic diversity is considered a good thing as it is a kind of insurance policy or future-proofing, increasing the chance that a species will have the right kind of genes to be able to adapt to change, such as a changing climate, or cope with a new disease. At the time of writing, the Takahē Recovery Programme is on the verge of getting detailed genomic information about the population (see Chapter 26), but until now Glen's genetic management has relied on using the takahē pedigree.

At its simplest, the pedigree is an enormous family tree that records – if known – the birth and death of every takahē in the managed population; their dam, sire and siblings; and the identity of any mates and offspring. There are plenty of birds from the 30 years of hand-rearing that started life as eggs or chicks in the Murchison Mountains whose parents are unidentified, and there are also plenty of birds released back into the wild whose subsequent breeding success has gone unrecorded. While it contains a few incomplete entries for wild birds in the Murchison Mountains, this population is essentially a genetic black hole. But for all the birds on islands and in sanctuaries around the country, the pedigree is an excellent record of a bird's life, and most importantly, of its relationships.

The takahē pedigree goes back to the start of captive rearing in Te Anau and then Burwood in the early 1980s, although importantly it also includes seven birds from early captive-rearing efforts at Pukaha / Mt Bruce in the 1960s and 1970s. The first entry in the pedigree is a male called Jock dating back to 1967, while number two is the female Red-Left, dating from 1972; both were collected from the Murchison Mountains and played a pivotal role in the development of today's takahē population. At the time of writing, the pedigree contains the records of 1,212 takahē, including 76 young birds from the 2021/22 breeding season. Just as racehorses share a birthday, so takahē all share a nominal first birthday on 1 October each year, which is when surviving yearlings are officially added to the population total. So the 76 youngsters were considered to be bona fide members of the takahē population only when 1 October 2022 ticked over.

Glen says he takes two important considerations into account for genetic management: the inbreeding coefficient and mean kinship. The inbreeding coefficient is a way of gauging how genetically close two birds are to each other. Full siblings have a coefficient of 0.5. If they mated and produced a chick, its coefficient would be 0.25. Each of our genes is made of two copies known as alleles, and we inherit one allele from each parent. The alleles can be the same or different, and if one allele carries a harmful mutation its effect will often be masked by the other. What the inbreeding coefficient indicates is the probability that an animal receives the same allele from both parents because they are related, which for the offspring of full siblings is 0.25, or a 25 per cent chance. The higher the inbreeding coefficient, the greater the likelihood that an animal inherits two copies of a damaging allele, which may allow its harmful effects to become apparent. The risk of small populations with high inbreeding coefficients is that you get inbreeding depression (see box), which often results in poor fertility – as seen in the kākāpō population. Another consequence can be reduced survival of offspring.

To avoid these deleterious effects, breeders try to keep the inbreeding coefficient low – for example, dog breeders aim to keep it below 0.05 (5 per cent). While there are some individual takahē with high inbreeding coefficients of 0.25 (and one very inbred Tiritiri bird achieved an eye-watering 0.344), Glen says that decisive management action in the past decade has seen the takahē population dodge the inbreeding bullet, as the average inbreeding coefficient is down from 0.0523 in 2012 to 0.034 in 2022, and infertility hasn't been an issue. In Glen's words, 'We got there just in time.'[1]

Glen's other important yardstick is mean kinship, which is essentially a measure of how related each bird is to all the other birds in the population. When it comes to mean kinship, the closer the number is to zero, the better. Tumbles, with no known relatives, was assumed to have a mean kinship of 0.00, which is why he was ranked so highly. All other founder birds from the wild Fiordland population are also assigned 0.00. The higher the number, the more common a bird's genes are and the more of its DNA is already shared with other birds. Glen says that the average for the takahē population is now about 0.02, which is a significant improvement from the 0.033 mean kinship measure in 2014.

The highest mean kinship value in the pedigree belongs to a deceased female called Iti. She lived from 1989 until 2006 and was the daughter of the prolific Squeak and Taku (see Chapter 14), so it's not hard to see why she was related to so many birds in the pedigree. With 10 offspring of her own, Iti not only continued in the family tradition of prolific breeding, but also of mating with family members. We've already heard how Squeak first paired up and produced one daughter with Tussock. A long relationship with Taku, followed which produced 11 chicks, the youngest of whom was Montague. After Taku, Squeak got involved in a threesome with her grandson Aho Ake (Iti's son) and a male called Bargie, and had two chicks of uncertain paternity. Her final suitor – and father of her last chick – was her own son, Montague.

But that was not the end of the incest. Montague then partnered up with his sister Iti, and they produced a female called Paku, who features in the Orokonui story (see Chapter 31). Paku's mean kinship value is almost as high as her mother's, which is why she was retired to Orokonui. And her own offspring also have high mean kinship values, sharing lots of common genes, and so are not priorities for breeding. This is why Paku's daughter Kawa found herself partnered with the unfortunately infertile Tumbles.

Iti's grandparents, meanwhile, sit at the beginning of the pedigree – Squeak was the daughter of Snow and Alpine, who were among the first birds hand-reared in Te Anau, while Taku is the son of Red-Left and an unnamed male at Pukaha / Mt Bruce. Around a third of all takahē in the pedigree are descended from a handful of these early birds, especially Jock, Red-Left, Shirley and Ralph from Pukaha / Mt Bruce, and Snow, Alpine, Mrs White, Mr White, Mrs Black and Mr Black from Te Anau. Most of these birds were relocated to islands, where they were left unchaperoned and the genetic mayhem took place.

A threesome on Tiritiri Matangi with their two chicks. Large family groups like this used to make genetic managment on islands difficult. MARTIN SANDERS

At the time of writing, the oldest birds currently represented in the pedigree are the ambassador males Orbell, at Zealandia, and Quammen, at Orokonui. The oldest female is Maroro, at Willowbank. All three have high mean kinships, meaning their DNA is already well represented in the population. The pedigree is a living document, and the mean kinship ranking – which orders birds in terms of genetic priority in the population – resets every time a bird dies. This is because mean kinships include only the living population, and so are recalculated and change when birds die and new birds are added. A bird's mean kinship goes up if it or its relatives produce offspring, because there are now more of its genes in the population.

One way Glen can calibrate the success of intensive genetic management is using genetic diversity, which is a well-recognised measure for managing populations. It calculates how many alleles have been 'captured' in the population and how well spread or represented they are. For each individual bird, the software models the likelihood of genetic diversity loss for the overall population if that individual dies during the year, taking into account its sex, age and relative rarity. The death of a rare bird has a greater impact on the population's genetic diversity than the loss of a common bird.

This is another measure showing that, despite everything, takahē seem to be in surprisingly healthy genetic shape. Glen says that the population still retains a good representation of the genetic diversity that was present when intensive management started in the early 1980s, and there is no evidence that inbreeding has had any deleterious effects, such as infertility. Conservation breeding programmes routinely aim to retain 90 per cent of source gene diversity. While takahē have 96.7 per cent diversity, in 2014 an expert panel agreed it would be possible to improve that to 98.6 per cent. Glen is quick to point out that we don't know what genetic diversity was lost prior to the 1980s, but he's hoping that some genome sequencing of old museum specimens will shed light on this.

Another way to measure success is founder genome equivalents. This is the number of unrelated founder birds from the Murchison Mountains you'd need if you were creating a new managed population of takahē from scratch and were trying to achieve the same genetic diversity as that held by the current population. In 2014, the actual founder genome equivalent for the managed takahē population was just over 15 birds, even though the real founder population was much larger that. The power of this measure is that you can use it to calculate the level of improvement intensive genetic management could create. For takahē, Glen has calculated that an achievable potential founder genome equivalent could be as high as 35 birds. This is reassuring, as the bigger the founder population, the better.

Overall, Glen recognises 43 founding lines in the managed takahē population. Each founder was an egg or young chick collected in the Murchison Mountains for hand-rearing, and Glen admits he doesn't know how they relate to one another. Breeding success has been very uneven; we've already seen how some birds have produced large, widespread family trees while others have only stunted family shrubs or have died with no living descendants. Back in 2012, says Glen, 90 per cent of the managed population came from just 10 of these family lines, but within five years good progress had been made to start to even out this lopsided representation.

The trick to avoiding inbreeding, minimising genetic drift (which is the gradual random loss of rare alleles) and maximising genetic diversity is to increase a population as quickly as possible, and at the same time purge common lines and boost rare lines by partnering rare birds with rare birds. The corollary of this is that common birds are partnered with common birds; this prevents common genes watering down the frequency of desirable rare genes, which happens with rare–common pairings. This involved a fair bit of matchmaking on Glen's part as he arranged the best genetic pairings between rare birds and ensured they were in the most advantageous location – usually at Burwood, at least to begin with. The idea was simply to produce as many rare chicks as possible. The Burwood takahē rangers monitored the pairs closely, taking first and sometimes even second clutches of eggs away to be foster-reared so that these valuable couples would be encouraged to lay again.

Glen has used several software programs to help in his analyses. These are widely used by many organisations that manage rare species, including zoos, and allow him to manage data, predict outcomes, simulate risks and compare various strategies. Until 2020,

Lost opportunities

Blaze was a Murchison Mountains founder with a very desirable mean kinship of 0.00. However, he lived only seven years and produced no offspring.

Hebe was a different kind of lost genetic opportunity. By the time of her death in 2015, this genetically valuable bird had, in fact, been very productive, with 11 chicks. However, all 11 were released into the Murchison Mountains, so she is not genetically represented in the managed population. Of course, it's entirely likely that her DNA still exists in the wild in Fiordland, and Glen crosses his fingers that that's the case.

the pedigree was recorded in a Single Population Analysis and Records Keeping System (SPARKS) studbook, and alongside this Glen used PMx software to manage the population genetics. SPARKS has since been replaced by the Zoological Information Management System (ZIMS), which is a global online database used by many zoos and wildlife parks, including places in New Zealand that are home to takahē, such as Auckland Zoo.

Glen was keen to ensure the takahē families each had representatives in at least one managed location in addition to Burwood, and in 2014, when the expert group convened to discuss the best genetic strategy for takahē, there were three birds whose lineages were not represented on the northern islands. Tumbles had no offspring, so nothing could be done about that, and Blaze died around this time, which took him out of the equation. That left the female Larivee.

Larivee was in the penultimate cohort of Burwood hand-reared birds in 2009 and is one of Glen's rare, valuable birds. Having wild Murchison Mountains parents ensured she had a mean kinship of 0.00 to start with – this has risen now she has some offspring, but not by much, as she is not a great breeder. To make up for that, Glen is now working hard to ensure her offspring are as productive as possible. She was partnered with Elwyn (grandson of Greg, and hatched and raised on Tiritiri Matangi Island) and their first chick was a female named Silberhorn, hatched in 2013; she remains at Burwood and has gone on to produce four living chicks. Chick number two for Larivee and Elwyn was Earland in 2014; she was released into the Murchison Mountains, where her reproductive successes, if any, aren't measured. Larivee was then partnered with Aparima and produced a son, Hossack, in 2016. He remains at Burwood and already has nine offspring. Larivee's final chick was

Mackenzie, hatched in 2017, who has produced one chick at Burwood. As their fathers had mean kinship values above 0.02, Larivee's offspring all have mean kinships around 0.015, which ensures each has a high ranking that keeps them on Glen's radar. Larivee herself is now considered retired and resides at Wairakei, on a golf course that boasts a predator-free fence and no shortage of grass. Relocated as a couple with Aparima in 2019, she has since paired up with Grant, while now former partner Aparima is with Miharo.

Another takahē whose rare family line has been front of mind for Glen is Arnie, a female from the same 2009 puppet-reared cohort at Burwood as Larivee. Her mean kinship is 0.01, putting her in the top 25 most valuable females. Like Larivee, she has been a slow breeder, although sometimes slow and steady is all that's needed. She and Larivee also have an ex in common, Elwyn, who fitted three partners into his seven-year life. Arnie shows how the best genetic planning in the world doesn't always produce great results, or at least not straight away. Arnie and Elwyn's 2012 chick didn't survive very long, and while her second chick, fathered by Taramea, lived nearly five years, she didn't breed.

In 2016, Arnie and Taramea raised the male Molesworth, who was moved from Burwood to Mana Island, which is a prime site consistently producing good numbers of chicks. Molesworth has got off to a promising start – by 2019, with partner Cape, he'd already fathered two chicks, Wairau and Te Rerehua, who have, respectively, been moved to Rotoroa Island in the Hauraki Gulf / Tīkapa Moana and Gouland Downs in Kahurangi National Park. Most recently, Molesworth has become entangled in a large breeding group on Mana Island that includes the male Swifty and females Cape and Matakitaki, and together they produced one chick in 2020. It was subsequently determined that this female, Tahu, is the progeny of Molesworth and Cape. In 2021, Molesworth and his three adult collaborators raised a further chick whose parentage was undetermined at the time of writing.

Meanwhile, Arnie and her new partner, Scotty, also seem to be getting in the swing of things, producing a daughter, Kohika, in 2019 and two sons, Kaplin and Franklin, in 2020. Arnie and Scotty's genes, and those of their parents and grandparents before them, have now been immortalised through the 2019 female chick Kohika, as she was chosen to have her genome sequenced on behalf of the whole species.

How big is big enough?

How many takahē does it take to make a healthy population? For many decades, and for many species, a population of 500 has been touted as the magic number to aim for. This is indeed what the first takahē recovery plan, published in 1994, put down as a goal. The latest plan has a revised goal: 'there are 250–1,000 mature individuals and the population growth rate is greater than 5% per annum over the period of the plan'.[2] Once this target is reached, then, all going well, the intention is that intensive management can be scaled back – or even stopped.

From a long-term genetic point of view, experts have calculated that a takahē population with a minimum of 106–120 breeding birds should be enough to avoid inbreeding depression. This is another consequence of close relatives breeding with one another, and it increases the chance that two birds carrying the same 'bad' or deleterious gene have offspring that then inherit two copies of this gene. A bird carrying one bad recessive copy is usually fine, as a dominant good gene will mask the bad gene. But two recessive copies generally means the disease or health condition caused by the gene is activated (or expressed, as geneticists put it).

Of course, more is always better, and the experts calculate that a population of 1,060–1,220 breeding birds would allow takahē to not only avoid inbreeding, but also to gain new genetic diversity as naturally occurring mutations occur, some of which might turn out to be evolutionarily advantageous in future.

At the moment, the takahē population finds itself in a genetically good space, and with 440 birds of breeding age and a new crop of young birds, it far exceeds the number required to avoid inbreeding (even if a good proportion of the birds aren't having any or many offspring). Next, Glen has his eye on getting to that magic 1,000-plus mark, but that then begs the question – where will all those takahē live?

CHAPTER TWENTY-SIX

TAKAHĒ GENOME

Kohika means 'ancestor', which is a very appropriate name for a takahē whose DNA has been used as the high-quality reference genome for the species. This particular female hatched in late 2019 at Burwood and is the daughter of Arnie and Scotty, both well-known birds when it comes to the takahē pedigree.

Kohika is part of the takahē genomics project, which was led by German researcher Lara Urban while she was a research fellow at the University of Otago. In February 2020, Lara joined the takahē team at Burwood to collect a fresh sample of blood from Kohika and both her parents. The team had deliberately chosen birds at Burwood as this made collecting and storing high-quality samples much easier than if the birds had been somewhere remote, such as in the Murchison Mountains. As soon as the small samples were collected,they were stored in liquid nitrogen and taken back to Dunedin for Lara to process in the lab and extract the DNA. Although the genome being sequenced is Kohika's, her parents' DNA provides a benchmark and contributes to the accuracy of the sequence. The DNA was then sent to the Vertebrate Genome Laboratory at Rockefeller University in the United States to be sequenced. This laboratory is highly experienced at sequencing reference genomes to the chromosome level, which is a higher level of detail than your average genome sequence. Because of the significance of takahē as a taonga species to Ngāi Tahu, it was important that the iwi gave approval for Kohika's DNA sample to be sequenced overseas.

Kohika is a female, which means both sex chromosomes are captured in the genome – avian sex chromosomes are Z and W, and it is the female that is ZW, while the male is ZZ (in humans, it is the male who carries a copy of the two human sex chromosomes, X and Y, while the female is XX).

A reference genome is the genetic equivalent of a type specimen in taxonomy. The type specimen is the single individual used as the basis for the description of a species, and it is the standard reference researchers refer back to if they have any questions about

Kohika's DNA has provided the high-quality reference genome for takahē. **LARA URBAN**

the diagnostic features of that species. A reference genome details how a genome – an individual's complete genetic blueprint – is structured and organised. Geneticists talk about the architecture of a genome and you can think of it as being akin to detailed architectural blueprints for, say, a complex house. These are very comprehensive blueprints that include all the minute design details of the house (the chromosomes), such as the curtain fabric and style of plates, as well as the basic structural elements such as walls and roofs. Any subsequent takahē genome sequences can now be mapped against the reference genome, and genetic differences easily noted (the house structure is the same, but there are different curtain fabrics and crockery choices, for example).

Kohika's is not the only takahē genome Lara was involved with. She was given access to a further 84 takahē blood samples that had been collected as part of routine health checks. The DNA from these samples was sequenced in New Zealand by the Otago Genomics Facility and the genomes have been referenced back to Kohika's genome. Although the cost of sequencing genomes has come down from millions of dollars in the case of the original human genome project to thousands of dollars, it wasn't possible or affordable to sequence the genome of every takahē, so the 84 were chosen to ensure that all the families on the pedigree were represented. This contrasts with the kākāpō, where the genome of every bird alive at the time was recently sequenced.

The takahē team hopes that all this detailed genetic information will help them better understand issues such as relatedness, and the levels and consequences of inbreeding. Glen has a fair idea from the pedigree about direct family trees, but the DNA information should confirm and greatly extend this knowledge, as well as enable a much more detailed look at how genetically diverse different individuals are. The hope is that researchers will then be able to work out if more inbred individuals produce fewer chicks, for example, or experience more illness. It might also be possible to combine genetic information with known breeding success from the pedigree to identify key reproductive genes.

Although it is very early days, the genome work has already clarified that takahē are in pretty good shape genetically, despite not having been very abundant since the last glaciation. And preliminary data analyses hint that those individuals that are least inbred are indeed having better reproductive success than birds that are most inbred.

Glen is particularly keen to find out if there are wild birds in the Murchison Mountains whose genes are not represented in the managed population, but he wants to do this without catching and handling birds. Lara has investigated whether it might be possible to genetically sample wild birds remotely by analysing DNA in droppings, which are abundant and easy to find. This work is just at the proof-of-concept stage, but Lara reports that she has managed to detect kākāpō DNA in poo, which is an exciting first step, and she says there is no reason the technique shouldn't also work for takahē poo. Lara believes that in future faecal sequencing could hold lots of potential for monitoring at arm's length, which will become more important as new large wild takahē populations become established.

CHAPTER TWENTY-SEVEN

BURWOOD – TAKAHĒ FARM

The neighbours' shrieks wake me at first light. I stumble out of bed, drag on some warm clothes and head outside to enjoy the riotous dawn chorus. I have spent the night at the Burwood Takahē Centre and my vocal, quarrelsome neighbours are several pairs of takahē living in pens near the office. They had serenaded me in the fading light of the previous afternoon, a counterpoint of shrill duets as they asserted their authority over their territories – not that there was any chance of a territorial takeover here, as each pair's territory is firmly demarcated by a fence. But in the fenceless wild, these displays are a way of ensuring that each pair knows its place both physically and socially. Takahē prefer not to see their neighbours, but a regular shout-out lets them know who's at home. The trumpet volley reminds me of their cousin the weka – the rowdy neighbour gene clearly runs in the rail family. After a few minutes, honour is obviously satisfied and the noise subsides. I'm pretty confident that all the birds concerned are now getting on with a takahē's favourite activity – eating.

****

The twice-daily takahē shout-outs are just one of many daily routines at what is fondly called the 'takahē farm'. Three days a week, the two resident rangers are on security duty, checking fences and traplines; the other two working days are food days, when each breeding pair of takahē gets a top-up of supplementary pellets. Today is a feed-out day, and ranger Sam Haultain and I get into the vehicle to head across the road to the breeding pens, where each pair of takahē gets their own plot of 1–1.5 hectares of copper tussock, rank grass and privacy. Sam has already sorted out the food – specially formulated takahē pellets made to order at Massey University. The takahē mostly eat tussock and grass; the pellets are more

A crowd of subadult takahē reacy to be paired up with their optimal genetic match.
SAM HAULTAIN

like a nutritious dessert that also ensures the birds don't eat themselves out of house and home. Sam explains that we'll quietly visit each pen, do the 'housework', and try to cast an eye over every takahē to ensure they're all healthy.

We park next to the perimeter predator-proof fence, which is designed to keep out feral cats, stoats and ferrets – and keep any errant takahē inside. There are several generations of predator-proof fence in use at Burwood, and they look a little different from the high, capped fences around sanctuaries such as Zealandia and Orokonui. The original, low Burwood fence is more of a DIY predator-resistant affair, designed in-house and erected by takahē staff. It is dug well into the ground to prevent anything excavating beneath it. A layer of wire mesh supports a layer of thick plastic, which has the double benefit of being slippery for anything trying to scale it and a visual barrier for takahē. Several electrified wires are an added deterrent for unwanted invaders. Because the fence is 'leaky', there are also trap boxes positioned on both sides of the fence to try to catch any predators that are either lurking outside or have managed to sneak inside. This fence design is still in use around some pens, but has been superseded by a higher, more effective fence that keeps out rats in addition to larger predators. This new fence has an additional high sheet of plastic, which means it doesn't need electric wires.

There are no gates, as these create weak points. Instead, there is an ingenious system of ladders, which we raise and then lower behind us. It feels a bit like raising and lowering a drawbridge over a moat. Each pen within the outer fence has its own mini-me fence, designed to keep each takahē pair safely in its own pen and out of sight of the neighbours. These fences are also regularly checked, as takahē are very good at finding weak points and breaking out to create havoc with the birds next door.

Burwood can be bitterly cold and windswept, but today is a warm, sunny late-spring day with just a gentle breeze to disturb the tall tussocks. The dry, papery rustle is like a gentle snare drum accompaniment to the loud chorus of skylarks ascending overhead. There are metallic glints as basking skinks scatter away from us. Sam points out the occasional takahē in the distance, but invariably, as we get close, they have all completed the magical takahē disappearing act in which a large blue bird with an ostentatious red beak that shouts 'look at me' has become invisible.

There is the same routine to follow in each pen. Keeping an eye out for glimpses of birds, Sam first cleans the shallow circular pool, which is like a concrete satellite dish sunk into the ground. With a large broom, she brushes it clean and swooshes the dirty water out onto the grass before refilling it from a hose. The pools are both drinking fountain and bath, and the state of them suggests they are well used.

Then Sam heads to the feeding station, which comprises two stakes secured in the ground and with brackets on top, into which is fitted a lidded hopper. The lid keeps rain and other birds away from the pellets, and takahē soon learn to flick it up using their head. Hygiene is critical, so Sam empties out any old pellets and cleans everything with an antiseptic wipe before refilling the hopper. The entire feeding station is also regularly relocated in the pen, as takahē never stop pooping and it doesn't take long for a pile of droppings to accumulate in the hopper's vicinity. This increases the risk of infection with coccidia, protozoan parasites that cause gastrointestinal issues.

Sam rattles the pellets enticingly as she empties them in, and noisily raises and drops the lid of the hopper to serve as a dinner bell. Then we back away a few metres to give the birds space, and crouch down to wait and see who turns up. Sam tries to read the combination of coloured bands on each bird's leg so she can identify it, and takes note of where they're at in the breeding cycle. It's November when I visit, part way through the takahē breeding season, and pairs are at various stages of breeding. If only one bird appears for a feed, it's a strong hint that the other might be sitting on a nest. If both birds are shy, they might be shepherding a very young chick. As chicks get older, they will accompany their parents to the feeder. Some of the families are actually trios, with a yearling helper.

A takahē slowly drifts towards the hopper, keeping an eye on us at the same time, and we catch a glimpse of its shyer partner lurking behind a tussock. Between the two there

Ranger Sam Haultain monitoring a takahē family at their feeding station, Burwood Takahē Centre. **ALISON BALLANCE**

is a constant reassuring refrain of quiet, deep belly thumps and high-pitched kweeps. The lead takahē reaches the hopper and with a confident flick of its head opens the lid before delicately pecking at the pellets. Sam says there is a range of feeding styles: she recognises peckers versus scoopers versus messy feeders who spill everything on the ground first. Unable to wait any longer, its partner breaks cover to join it at the hopper. This one is definitely more of a messy eater.

As we move between pens, it's interesting to observe the different personalities of the birds. Some are shy and cautious, others are quietly curious, some are just obsessed with food, and yet others are fearless and even friendly. Sam and other rangers over the years have had many varied takahē encounters – they've been avoided, followed, and sometimes even intimidated by bolder birds. They have experienced the thrill of seeing newly hatched chicks for the first time, and of seeing them grow into gawky adolescents and finally big, glossy adults. They've also experienced the disappointment of infertile eggs and the sadness of deaths. Sometimes, there is just an unexplained disappearance. They've witnessed immature island birds learning how to feed on tussock and developing the skills required to dig for *Hypolepis* in the forest pens, before heading back to the species' ancestral home in the Murchisons. The rangers get physically close to the takahē on a daily basis, but they never forget that these are wild birds, that working with them is a rare privilege, and that the successes of today are built on a foundation of decades of work by those who came before.

I visited Burwood Takahē Centre in late spring 2018, seven years after it became the 'engine room' of the Takahē Recovery Programme. It achieved this by changing its emphasis from captive rearing to captive breeding – gone were the intensive days of incubating eggs and puppet-feeding chicks; resident pairs of birds would now raise their own families. This would meet the programme's short-term goal of growing numbers quickly, as a key part of genetic rescue for a species is playing a numbers game. So what did this all mean on the ground?

Post-2011, the captive breeding operation was significantly expanded, from five to 20 resident pairs of takahē. Burwood also boosted its role as Grand Central Station for the whole Takahē Recovery Programme, with birds coming and going at all times of the year and staying for various durations. From just 30 or so birds during the 2000s, there were now up to 100 takahē at Burwood in any one year. In addition to the breeding residents, there are young birds from the previous year, island birds in transition to the Murchison Mountains or Gouland Downs, or off to a new sanctuary, and subadults or adults going through the takahē version of speed-dating.

Lots of birds meant that lots of new pens had to be built. These come in different sizes but all include a predator-proof fence. There are good-sized breeding pens, training pens that include beech forest as well as tussock, large pens for groups of yearlings, quarantine pens for new arrivals or birds in transit, and forced mating pens where genetically desirable matches are made.

All this makes for a busy year. It begins in the depths of winter – young birds are busy learning about tussock, *Hypolepis* and snow, and the team is busy behind the scenes checking that all the takahē pairs that are needed for breeding are in place, and that all the pens have solid fences. This will be important when hormones surge in spring and the usually peaceable birds start itching for a fight. Supervising ranger Nichollette Brown says the birds get very determined about breaking out. By early spring, it's time to leave the birds alone to get on with courtship and egg-laying undisturbed, apart from the feed-out rangers doing the twice-weekly roll call.

The rangers note presence and absence – and if they don't see the female for two weeks, it almost certainly means she's sitting on eggs. Female takahē usually incubate during the day, with the male taking the night shift. However, this regular pattern goes out the window if the pair have a yearling helper. The eggs are very delicate during the first few days of incubation, so it's important to ensure they are not disturbed until they are at least eight to 10 days old.

Once the two-week absence has passed, it's time to do a check. For many years, that was done using a takahē dog that could sniff out the nest quickly and with minimal fuss and noise. That dog was either long-time ranger Chris Phillips' dog Pai or Glen Greaves' dog

A young takahē undergoing winter *Hypolepis* training in the Burwood beech forest.
JAMES BOHAN

Yuki, who enjoyed a sedate walk in a pen even when she was well past her days running around the Murchisons. Now that both dogs have retired from takahē duty, all the female birds are fitted with radio transmitters and found that way. The ranger marks the nest location and makes a note of how large the nest latrine is – the incubating bird has to answer the call of nature frequently but doesn't want to go far, so the size of the pile of droppings is a good indication of how far advanced the incubation is.

All this information is written up on a nest board in the office, which gives the rangers an overview of where all the birds are up to in terms of nesting. Then, when the eggs are judged to be 10–14 days old, it's time for their fertility to be checked. Using a special candling torch to illuminate the size of the air sac, the development and position of blood vessels, and the developing embryo, the observer can see whether the egg is a dud or a winner and accurately calculate its age. About 80 per cent of takahē eggs are fertile, which is reassuringly high compared to many other endangered birds.

Nichollette and the team now have to decide what happens with any fertile eggs. Do they stay with the natural parents? If the pair is genetically valuable and it's early in the breeding season, they might remove the eggs to encourage the birds to re-nest and hopefully lay more. The fertile eggs will be fostered to another pair. These might be birds that have infertile eggs or have just a single fertile egg. The challenge is finding a match with another nest at the right stage. Takahē usually raise just one or two chicks, but the team has recently

Miharo

Miharo is a recordholder in the takahē world, having produced the greatest number of chicks of any individual: a mighty 21 (although she has outlived a few of them). She started life as an egg in 2005 in the Murchison Mountains and was hand-reared at Burwood, where she has spent almost her whole life.

Her first mate, Ananda, was six years older than her and they produced four offspring. Her second mate, Nuinea, was three years her junior, and together they produced 16 offspring, which is quite the legacy. These birds are scattered across many of the current takahē sites, as well as at Burwood. Both Miharo and Nuinea are alive at the time of writing, but are no longer together. Nuinea remains at Burwood, while Miharo has been retired to Wairakei Golf + Sanctuary. Here, she is going strong with her third mate, Aparima; in 2019 they produced the male Sinclair, who was part of a group of young birds moved to Gouland Downs in 2021.

carried out a small trial to see how three-egg clutches might succeed. Results from year one were mixed: one of Wairoa and Pahi's eggs didn't hatch, one chick was lost early on, and the third chick was successful. But George and Wall successfully hatched and raised three large chicks, showing that it is possible, which gives the team more options in future.

Any infertile eggs are removed, and if all the eggs in a clutch are infertile the nest is destroyed, again to encourage re-nesting. After these early nest checks, the birds are left on their own to get on with the business of hatching and raising chicks. The regular ranger makes food deliveries, does roll call and keeps the nest board up to date with births and any deaths, but there are no other visitors until after Christmas, by which time most chicks have hatched and are growing fast.

Parents start taking their chicks to feed at the hoppers when they're a month or so old. At this stage the lid of the hopper is bolted open, as a small chick isn't big or strong enough to lift it. There are no kiddies' meals for takahē – they are allowed full portions of exactly the same pellets as the adults eat.

Activity ramps up in February and March, when the chicks are caught for the first time so they can be banded and have a full health check. The chicks get their initial vaccinations, and their parents get booster shots as well as an annual health check.

Late summer often sees new arrivals at Burwood. Sometimes, chicks of the season from

Nichollette Brown with a takahē chick being raised by its parents at Burwood. DOC

other sites arrive for fostering. Nichollette says there is a real art to successful fostering, and playing safe and conservative is the order of the day. The team know the personalities of the foster pairs and keep a very close eye on the birds as the process is underway to prevent serious issues. One important factor in choosing which chicks to keep together as a group is weight – they need to be about the same size, so there are no big ones to bully little ones. Some aggression is inevitable and is critical to establishing strong family bonds, teaching chicks respect for their elders and helping them understand their place in the hierarchy. But even so, a ranger maintains a close watch for the first 12 hours, keeping an eye out for serious issues such as injury or exhaustion. The birds are kept in a small pen with solid walls, so there are no distractions and their focus is solely on one another. They are then observed for another few days to ensure that good relationships are established.

Around the same time, there is often an influx of carefully selected subadults to be paired up with an optimal genetic match. Matchmaking takahē style is known as 'forced pairing' and simply involves putting the male and female together in a small enclosure. If that first introduction goes well and the birds take a positive interest in each other, then after a few days the new couple are released together into a large holding pen. Takahē are very accommodating about having their mates chosen for them and such pairings are generally very successful. These new pairs will be moved out to various sanctuary sites or remain at Burwood to replace existing breeding pairs that have had a successful time breeding and have reached their genetic quota. Most pairs breed for five to six years before being moved on to another site.

The next batch of new arrivals is usually in May. These are semi-independent juveniles of the year arriving from various sanctuaries for Fiordland survival training before release in the Murchisons the following spring, after the winter snow has mostly melted. They're fostered in groups to trainer pairs, which introduce them to *Hypolepis* and teach them grubbing skills, and they'll acclimatise to the cold Southland winter. Among the young birds heading into the hills are some of those born and raised at Burwood – with two-chick families, one will often be released while the other will remain as a helper for the next breeding season.

The final move of the year might be in July and August, when seven- to eight-month-old birds arrive from grassy sanctuary sites to attend tussock training school. The latest iteration of this has seen a move away from adult teachers to teaming birds with similar-aged buddies that are already experienced with tussock and readily pass this know-how on to their peers. These birds will be moved to Gouland Downs or to some of the wilder sanctuary sites.

The annual process of deciding which birds are going where is a big chess game of possible genetic outcomes. These then have to be married to the logistics of moving birds from remote islands to equally difficult destinations all around the country. Planning involves the entire team: Glen decides on the best genetic matches, James Bohan allocates space at Burwood, Phil Marsh knows the state of play at all the sanctuary sites, Jason van de Wetering knows what's happening at the wild recovery sites in the Murchison Mountains and at Gouland Downs, and Nichollette oversees everything to make sure it is resourced properly. Plans are made and travel is booked – it's a complex arrangement involving pre-travel health checks for birds, and then arranging for them to be caught on the right day and carried on foot or by quad bike to pick-up points for transfer to boats, helicopters or cars.

A typical Burwood year in numbers

On 30 September 2016, DOC announced that the takahē population had reached a milestone of 306 birds, the first time in many decades it had been that large and an increase of nearly 10 percent from the previous year. Just over a third of those birds were in the Murchison Mountains. Burwood had started the year with 74 birds in residence and ended up with 69 birds. During the 2015/16 breeding season, there were 19 Burwood breeding groups, which produced 44 eggs, three-quarters of them fertile. Twenty-eight chicks hatched, and following four deaths the season ended with 24 juveniles.

Eleven birds were imported to Burwood from various sanctuary sites to replace older breeding birds. And 37 birds left Burwood, with 29 of those being released in the Murchison Mountains. The other birds – four breeding pairs – headed to four different sanctuary sites.

From there, it's on to local airports, where the birds travel for free on scheduled Air New Zealand flights. People then need to meet the flights and uplift the birds for onward travel. And when the birds finally reach their destination, they need to be checked and provided a quiet place with food and water so they can recover and settle in. Once you remember that takahē are currently living in 20 places, from Foveaux Strait to north Auckland, and that many of these sites are not managed by DOC but by councils, sanctuaries, iwi groups and private individuals, the delicate logistical spider web becomes even more impressive.

But while a takahē move exemplifies people's ability to coordinate and work together, there are plenty of other factors that don't always cooperate with the grand plan. Bad weather and Covid-19 lockdowns can interrupt the best-laid travel itinerary, and so can the birds. If they decide, for example, to start breeding early, then that can disrupt spring moves – which is exactly what happened for the first transfer of birds to Gouland Downs in Kahurangi National Park.

CHAPTER TWENTY-EIGHT

MICROBIOME

In the past few years, researchers have become increasingly interested in the communities of microbes living in the digestive systems of animals. There are two main terms used: microbiota, which is an individual's collection of microorganisms; and microbiome, which is all those microbes, along with their collective genomes or DNA. The microbiota includes bacteria, viruses, fungi and archaea, and because these organisms are so small they're often identified just from genetic sequences, rather than from how they look. While omnivores such as humans can rely on enzymes to break down proteins and simple sugars, mammalian herbivores rely on microorganisms in their gut to break down fibrous plant material and convert complex molecules to more easily digestible short-chain fatty acids. These beneficial microbes in the gastrointestinal tract play an important role in nutrition, health and immunity. On the flip side, an imbalance of too many 'bad' microbes such as *Campylobacter* can cause gastrointestinal disturbances and ill health. Quite a lot is known about the microbiome of mammals, but researchers are only just turning their attention to birds.

When it comes to digesting its endless diet of tussock bases and pasture grass, it turns out that a takahē is somewhere between a giant panda and a cow. All three species eat large quantities of low-quality, fibre-rich plant material, but they have different ways of breaking down a plant's sugars, starches, cellulose and hemicellulose to extract energy. Cows eat a range of grasses and other leafy plants. They are ruminants with a specialised multi-chambered digestive system that uses microbes in the rumen to ferment food. This creates a soft cud, which the cow then chews a second time before it is swallowed and passes into the stomach for digestion to be completed. A diverse range of gut microbes is key to this efficient strategy.

Giant pandas, on the other hand, subsist strictly on bamboo. Although they are now plant eaters they still have the straight gut of their carnivorous ancestors. Unlike their meat-eating relatives, however, they have a very low diversity of gut microbes and are not particularly efficient at extracting nutrients. Their solution is simply to eat in bulk – which is pretty much what takahē do. But it turns out that despite the takahē eating a diet that is almost as monotonous as that of the giant panda, its microbiota is quite varied and actually very average for a herbivorous bird.

In a surprising twist, New Zealand's other surviving giant flightless herbivorous bird, the kākāpō – which eats a very varied diet – has a microbiome that is very low in diversity and, in that respect, is very similar to the giant panda.

In 2017, University of Auckland researcher Annie West began the first investigation of the takahē microbiome as part of her doctoral research. Using genetic techniques, she investigated bacteria in fresh takahē droppings, which gives insights into microbes living in the lower or hindgut. Her samples came from 57 birds living in eight sites, including wild Murchison Mountains birds and captive birds at Burwood, as she was interested in finding out whether being in captivity caused any change in their microbiomes compared to those of wild birds. Her main finding was that the dominant organisms of the takahē's moderately diverse gut fauna are members of three well-known bacterial phyla: Bacteroidota, Bacillota (formerly Firmicutes) and Pseudomonadota (formerly Proteobacteria). Six other phyla are also present, but in low concentrations.

The most abundant genetic sequence, which accounted for about a quarter of the 2,000-plus strains of microbe detected, comes from a genus of Bacillota called *Lactobacillus*. The Bacteriodata genus *Prevotella* was also common. *Lactobacillus* are common bacteria found in yoghurt, for example, and are often used as gut probiotics. The genetic sequence of the subspecies *Lactobacillus aviarius araffinosus* was the most common and present in nearly every sample. This particular *Lactobacillus* lives in anaerobic conditions and aids the fermentation of carbohydrates to produce lactic acid, which helps takahē get energy from their food.

Annie found that there was a wide variation in the composition of microbiomes between individual birds and also between sites. The most significant result was that, compared to the other sites, the samples from Burwood had the lowest microbial diversity and a much higher prevalence of *Lactobacillus*. Annie didn't know for sure, but suggested that the fact captive Burwood birds eat lots of cereal-based pellets and experience more medical treatment that other takahē populations may partly explain this difference.

Annie says it's early days, but she believes that in the future a deeper understanding of the takahē gut microbiome might help conservation managers keep the population in optimal health.

CHAPTER TWENTY-NINE

THE RETURN

We meet at eight in the morning at Burwood with a flurry of introductions. It's 70 years, almost to the day, since takahē were rediscovered, and the children of that discovery team have joined the Takahē Recovery Programme team to mark the occasion: Geoffrey Orbell's daughters Lynley and Mairi; Joan and Rex Watson's daughter Paula and son Huntly; and Neil McCrostie's son Dave, here with wife Helen. Takahē rangers Glen Greaves and Julie Harvey lead the way to a small pen by the original hand-rearing centre and explain how we'll catch the two takahē. The others watch on as Julie and Glen quietly herd the birds into a corner where they'll be easier to grab. The birds protest loudly and move quickly, but so do Glen and Julie, and they soon have a bird each. Holding a takahē is definitely a two-handed job – one hand firmly behind the head to control that strong, sharp beak, and the other holding both legs securely to prevent any kicks.

Not everyone in the group has seen a takahē close up before, and plenty of photos are taken and admiring remarks made. And the admiration is well deserved – takahē are handsome birds, and the bold red–blue–green colour scheme makes quite the fashion statement.

These two are a very young pair called Dore and Tauhou. One-year-old Dore is the daughter of Lily and Rerehu, a pair of Burwood breeders who were among the first birds moved to Kahurangi. One-year-old Tauhou was raised on an island in Foveaux Strait by his mother Celie and father Tangaroa. Both birds are below average in terms of genetic value, so are not needed at any of the sanctuary sites.

The birds are gently put into wooden travel boxes that are well supplied with air holes, and then loaded into cars for the drive into Te Anau. We are heading into the hills by helicopter, piloted by well-known identity Sir Richard 'Hannibal' Hayes, who says he's flown plenty of takahē in his time. Our destination is Tor Ridge in the Murchisons, just a few minutes' flight directly across the lake from the township and south of Takahē Valley. As we travel across

Descendants of the original rediscovery party release two takahē in the Murchison Mountains in 2018. **ALISON BALLANCE**

a glassy, calm lake, we can hear the two takahē calling inside their boxes, loud enough to carry over the engine noise. Fiordland is an impressive landscape on any day, but where we come into land is next-level spectacular: a high ridge covered in monumental weather-worn limestone spires, like a collection of creamy-grey Henry Moore sculptures.

Hannibal lands, and as the helicopter blades come to a halt, the loud engine noise gives way to the warm silence of a calm day in the alpine zone. A pipit calls in the distance and a couple of alpine cicadas rasp quietly. There are exclamations over the scenery and the expansive view back across the lake to Te Anau. The ground is moist and squelchy, a miniature garden of ground-hugging mosses, lichens and cushion plants.

Glen carefully lifts out the takahē carry boxes and places them on the ground. The family members gather where they have a good view of the boxes and get their cameras ready to record the moment the doors are raised. The two birds rush out, but after a few metres they pause, look around and take a few cautious steps. Then they break into a relaxed run, heading downhill, where they are soon obscured behind a limestone spire.

****

Dore and Tauhou were released on Tors Ridge in 2018 to mark 70 years since the rediscovery.
ALISON BALLANCE

This return took place in 2018, and it followed a tumultuous few years in the Murchison Mountains. Back in 2010, the takahē team decided that the wild population in the Murchison Mountains would be left pretty much to its own devices. There were to be no new arrivals from Burwood, but equally there would be no eggs or chicks leaving either. The previous two inbound transfers of 13 and 16 yearlings, in late 2008 and 2009, respectively, had helped the population recover to around 109 birds, and the social disruption caused by the loss of many mates during the 2007 stoat plague had subsided. There were now 26 pairs, a big increase on the 17 pairs in the year following the big loss.

Following the 2011 DOC reshuffle, the takahē team were no longer directly responsible for stoat and deer control in the Murchisons, although these efforts would continue as before, with the intention of providing protection for species such as Fiordland tokoeka, whio and kea, as well as takahē. It was a tough-love management decision, to stand back and see if the expanded trapping programme, which now covered 50,000 hectares, would give the birds enough security to ride out any future stoat plagues. It was also a pragmatic decision, to reallocate limited resources in the takahē programme to other priorities, including expanding Burwood and paying attention to neglected sanctuary sites such as the islands.

The Murchison birds weren't completely off the radar. There were now routine Sky Ranger flights to monitor a subset of 50 or so birds, and takahē rangers still dropped in to fit radio transmitters to new birds and replace ones whose batteries were running down. This monitoring took into account how many birds died, disappeared or lost their transmitters, and calculated a survival estimate and population size. A survival rate of 0.84 was the break-even mark – lower than that and the population would be in decline, and above that it would grow. For the first full year of this new system, the survival rate remained above the break-even mark and the population held steady.

But trouble was already brewing, as hot summer temperatures in January and February 2010 set in motion another drawn-out mast-seeding and stoat plague sequence. By spring 2010, about 65 per cent of the tussocks in Takahē Valley were flowering and the mountain beech was festooned in red flowers. That translated into an autumn 2011 beech seedfall over nearly 12,000 seeds per square metre. Mice and rat numbers shot up. And with a mild winter they stayed up – and up, right through the following year. With so many rodents to eat, stoat numbers began to climb, too. In February 2012, DOC staff decided to increase the frequency of trap checks from four times a year to six, and by autumn they had elected to increase that even more, to fortnightly. An extra 500 trap boxes were added to the network.

The winter of 2012 saw very high catch rates of rats and stoats in the traps. Then, in September, two years after the beech and tussock flowering lit the match on this latest slow-burning plague, rat numbers crashed. Now that the traps weren't full of dead rats, they began catching many more stoats, but this was just a reflection of how many starving stoats were out there – there were plenty still on the rampage. In the final months of 2012, the latest batch of stoats left the nest and began turning their attention to takahē. At least four takahē were killed by stoats, and perhaps more.

Then came the icing on the disaster cake – heavy rains in January 2013, which caused significant flooding. Five huts in the Murchisons had rivers flowing through them, and four takahē are known to have drowned. The Murchison Mountains takahē population experienced a mortality rate of 0.255, and finished the 2012/13 season with just 84 birds. This worrying decline came on top of further bad news – another mountain beech mast was kicking into gear, with bountiful spring flowering.

Heavy beech seeding followed in autumn 2014, and by spring the rat population was on the rise. At the same time, the Murchison Mountains takahē population declined to just 83 birds. This was not a viable population for the long term and alarm bells were ringing in takahē HQ. In anticipation of the coming stoat plague and to try to minimise the damage, 1,000 more traps were added to the network, bringing their numbers up to nearly 3,500. Meanwhile, the rodent fuse on the stoat time bomb ticked away – rat numbers peaked that winter and began to fall in spring.

Apiata – tale of a wanderer

Ranger Phil Marsh has met Apiata the takahē on quite a few occasions. And nearly every time, the peripatetic and lonely bird has been in a different place. Phil says he epitomises the distances takahē are willing to walk to find a mate, and is a favourite bird 'in light of how much of a survivor he has been'.[1]

Apiata was hatched on Tiritiri Matangi in late 2008. He is the son of Mahuika and Blackwatch, and great-grandson of the famous Greg, and was named by James, one of the island rangers, after Willie Apiata, who the previous year had become the first recipient of the Victoria Cross for New Zealand. When Apiata was four months old, he was sent to Burwood to learn tussock and *Hypolepis* skills from a pair of foster parents, and he was then released into the Murchison Mountains when he was a year old.

The first time Phil caught up with Apiata and his new partner was in 2011 in the Murchisons. The pair were west of the Top McKenzie Hut, and Phil and his colleague caught both takahē to change their radio transmitters. Apiata's partner died during 2012 and his transmitter fell off at about the same time. 'That is when his gypsy days began,' says Phil. In 2013, Phil managed to find Apiata, who was still single, and fit him with a new transmitter. After that, Apiata was on the move, looking for love.

During 2014, the Sky Ranger monitoring system detected Apiata far out to the west, outside the safety of the stoat trap network, and the team decided he should be relocated back into the trapped area. Phil recalls that the bird was in a spot on the slopes of Mt Irene where the helicopter couldn't land properly, so the rangers had to scramble out as the chopper hovered. The rangers knew from the radio signal that they were almost on top of Apiata, but they walked past him several times before they eventually located him, frozen to the spot and tucked under thick vegetation. Rather than carry him in the usual fashion, in a handheld cloth bag, they secured him in a backpack for the scramble back on board the helicopter and the flight to his new home in Plateau Creek.

Three years later, Phil was back in the Murchisons for the three-yearly ground survey when he came across Apiata – still a lonely bachelor – in a new location near Sheerwall Creek. On his next visit, Phil found Apiata in yet another new location – Point Burn – only this time he had a mate. Phil and his colleague eventually caught both birds, fitting them with radio transmitters held in place with new-style flight harnesses, but not before a lung-crushing pursuit up a steep slope.

In 2020, Phil and a colleague were back in the mountains for a day, catching birds and fitting them with transmitters to maintain a good sample size for Sky Ranger tracking. Apiata's radio transmitter had stopped working, but they managed to find him via his partner's radio signal and successfully fitted another transmitter. Phil doesn't know if Apiata has managed to become a father, but hopes that after a few years with the same partner he has finally had some success raising chicks.

Phil is hoping to catch up with his old takahē acquaintance soon. 'I'm expecting that I'll once again come across Apiata on my travels – it's just the way it is!'

James Bohan radio-tracking takahē in the Murchison Mountains.
LISA VAN BEEK

The young female takahē Hakapoua was raised at the Burwood Takahē Centre and released in the Dana Peaks area of the Murchison Mountains. Glen Greaves caught up with her in 2017 when she was two-years old and had relocated to Takahē Valley. **ALISON BALLANCE**

Every third year the takahē team ground-truth the Sky Ranger population estimates. In November and December 2014 survey teams visited every territory in the Murchisons and counted every adult takahē. Their tally was just 78. At the same time, stoat numbers went through the roof just as their abundant rat dinners disappeared. But the intense trapping seemed to work – by the end of September 2015, the takahē population remained steady at 77 birds. It was the lowest number of takahē ever recorded in the Murchison Mountains, but at least it hadn't plummeted any further – it was a far better outcome than any recorded in previous stoat plagues.

However, in yet another groundhog day repeat this was all coinciding with another beech mast in Takahē Valley, which began to drop seeds in autumn 2016. This would roll on into another stoat plague in the summer of 2017/18.

So why not use the toxin 1080 in the Murchisons to solve the problem? This is highly effective at controlling outbreaks of predators, but DOC has always erred on the side of caution and refrained from using any kind of toxic bait in the area, so that there is no chance of takahē eating these, or even eating poisoned mice and possibly feeding them to young chicks. And the takahē team was now much more confident that frequent checks of the widespread trapping network at key times, such as when stoats give birth in spring, would prevent a repeat of the 2007 stoat slaughter.

The takahē team also decided it was now in a position to do something to boost the ailing Murchison Mountains takahē population, and that came in the form of big influxes of new birds. The new arrivals were mostly surplus yearlings from the sanctuary sites and from Burwood, all parent-raised and suitably trained in the art of tussock and *Hypolepis* browsing. The goal for the next few years was to fill the Murchisons to its capacity of 200 birds, so that every territory was occupied by a pair. The first new tranche of 29 birds was released in November 2015, followed by other large annual releases: 26 birds in 2016, 30 birds in 2017, 20 birds in 2018 (including Dore and Tauhou), and what was intended to be a grand finale of 41 birds in 2019. At this point, the Takahē Recovery team announced they were hanging up the 'no vacancy' sign.

The first two releases, however, hit the ground just ahead of the incoming 2017/18 stoat storm. This time, the takahē team had an emergency plan in place – they would keep a close eye on the Sky Ranger numbers, and if modelling suggested high levels of predation, they could take two courses of action. In an average year, about 10–15 per cent of the Murchison Mountains takahē population dies. If the death rate rose to 25 per cent it would show that trapping alone wasn't enough to protect the birds, and it would trigger a critical review of the management and future of the Fiordland birds. If the death rate rose to 35 per cent, it would trigger the emergency evacuation of genetically valuable birds. But while some birds were still killed by stoats in 2017/18, the total was far fewer than in 2007. The death rate reached 20 per cent, which was unfortunate but not calamitous; it meant that the trapping scheme was effective, and the population now plateaued in a plague year rather than plummeted.

This was just as well, as the Murchison Mountains would face yet another mast year before the decade was out. This fourth event was a doozy, a country-wide mega-mast that started in 2018, gathered rodent momentum through 2019 and culminated in a tsunami of stoats in 2020. This mast event would also feature prominently in the story of Gouland Downs and efforts to create a new wild mainland takahē population there.

Takahē and 1080

In the summer of 2018/19, a mega-mast event was predicted to erupt across large swathes of New Zealand's forests and high country. An exuberant abundance of fruits and seeds, produced by countless tree species and by numerous different snow tussocks, would encourage many native bird species – such as orange-fronted parakeets and kākā – to breed. Unfortunately, the forecast was that mass seeding would also drive mice and rat populations to record levels in many places. This rodent boom would be followed, many months later, by a significant spike in stoat numbers, to such an extent that the usual trapping networks would probably be overwhelmed.

Tiakina Ngā Manu – Battle for Our Birds is DOC's nationwide predator control operation, and during the mega-mast the team intended to roll it out across a million hectares of land with high conservation values and vulnerable species. The operation uses 1080, which is highly effective: rats and mice are killed by eating toxic baits directly, and stoats die from eating secondary poisoning, after eating rodents with 1080 in their stomachs.

As part of this effort, DOC was planning to use aerial 1080 in the area around Gouland Downs in Kahurangi National Park. The aim was to keep it safe for the native locals – great spotted kiwi and giant *Powelliphanta* land snails – as well as newly introduced takahē. The core takahē area in Gouland Downs was to be excluded from the 1080 operation area as there was uncertainty over whether takahē might eat the 1080-laced cereal pellets – after all, many of the birds had been 'trained' to eat special takahē pellets, which might predispose them to trying the pellets containing toxin.

Before takahē were moved to Northwest Nelson, it wasn't known how takahē respond to 1080 if they ate the toxin, although it was thought that their large size would be protective in that they would have to consume a lot of pellets in order to receive a fatal dose. The closest analogue to how takahē might respond came from observations of their rail cousin, the weka. Weka that had accidentally eaten 1080 pellets had been observed appearing unwell and then recovering after a few days. Unfortunately, becoming ill didn't seem to teach weka to avoid similar pellets in future, and it was likely that takahē would behave similarly. So DOC carried out research into takahē repellents, smells and tastes that would naturally discourage the birds from even trying the pellets. Similar research was underway with kea, and the challenge lay in finding a repellent that didn't also deter the rats and mice that were the main target. Unfortunately, none of the proposed takahē repellents tried so far have worked very well.

Since an effective repellent is still a long way off, if it ever happens, the takahē team carried out feeding trials with takahē to assess the likelihood that a bird would eat a 1080 bait. The trials took place even before takahē were taken to Gouland Downs, using non-toxic pellets at two sites: Burwood, where birds received supplementary food in the form of pellets; and an island in Foveaux Strait, where they didn't. Footage from remote trail cameras showed Burwood birds eating lots of the non-poisonous pellets, but on the island the birds were completely uninterested, to the point that they would walk right over pellets as they focused on feeding on grass.

The takahē team was between a rock and a hard place. It desperately needed a new large mainland home for the growing takahē population, yet there was this unknown level of risk to the birds during the occasional mast year, when trapping alone wouldn't be enough to prevent takahē losses and 1080 couldn't be used to significantly knock back predator numbers. But they also knew the risk would only last a few years, until the Gouland Downs founder birds that had been at Burwood died. Their offspring wouldn't have had the same exposure to pelleted food and would blissfully ignore any baits, just as the island birds did in the trial. They might not even have to wait for all the founders to die: pellet trials were carried out at Gouland Downs just two years after the birds were released, when they stopped receiving supplementary food, and their level of interest in non-toxic pellets was already much more akin to the island birds than Burwood birds.

In the end, DOC's Tiakina Ngā Manu team, after discussion with the takahē team, decided that the best option would be to use 1080 across a large area of Kahurangi, but not in the core takahē area. With the plan in place, it was a matter of deciding the best time to apply the baits so that the operation was most effective at protecting nesting birds.

CHAPTER THIRTY

KAHURANGI – A NEW WILD HOME

In 2016, the takahē population passed 300. This meant that the species was no longer teetering on the brink of extinction, and DOC celebrated the milestone by moving it down two steps in the New Zealand Threat Classification System. Formerly classed as Nationally Critical, the takahē was now Nationally Vulnerable, and the long-discussed goal of creating a new self-sustaining mainland population became more urgent. The existing island populations were pretty much at capacity, and the steady release of Burwood birds back into the Murchisons would soon bring that habitat up to its estimated maximum carrying capacity of 200 birds. Glen Greaves remarked, 'We know how to make a lot of takahē – but we don't really have a home to put them.'[1]

The Takahē Recovery Team did have a possible place in mind, however: Gouland Downs, on the Heaphy Track in Northwest Nelson's Kahurangi National Park. It is estimated that this area could support perhaps 30–40 pairs of takahē and it was considered suitable as a new wild recovery site for a few reasons. For a start, takahē historically lived in the general area. Second, it undulates between 700 metres and 800 metres above sea level – high enough to get some snow in winter, but with a milder climate and lacking the avalanche-prone high peaks and passes of the Murchisons. And third, there was no need to build new infrastructure, as existing huts on the Heaphy Track – which is maintained by DOC as a Great Walk – could be used by the takahē team.

Gouland Downs is a large expanse of bog, tussock and subalpine scrub, surrounded by beech and broadleaved forest that acts as a wide buffer to the farmland to the east. There are plenty of suitable plants for takahē to eat and year-round water, although there is no *Hypolepis* for the winter months. Another black mark is that the soil has low fertility. The area is home to rats and stoats, but importantly seems to lack feral cats and ferrets, and

Gouland Downs, on the Heaphy Track. **DANILO HEGG**

the stoat and rat populations here have been controlled over many years. A further 500 predator traps had recently been added to the original network of 160, increasing the size of the trapped area tenfold to more than 5,000 hectares. This trapping is a collaboration between DOC, Air New Zealand and Manawhenua Ki Mohua iwi. There has also been regular use of aerial 1080, providing effective predator control at a large landscape scale.

With the lessons of the largely failed attempt to establish a second Fiordland population in the Stuart Mountains in mind (see Chapter 15), the takahē team knew what they needed to do: release a significant number of birds together, and have a good mix of older experienced birds alongside younger unproven birds. By August 2017, 30 birds had been chosen, some from Burwood and some from various islands. The latter had already spent their winter at Burwood, acclimatising to colder winter weather and learning to eat tussock instead of pasture grass. Everything was ready, but then a succession of bad weather kept delaying the release, and by the time it had improved, the birds were already showing signs of breeding. Rather than disrupt them, the team decided to postpone the move for a few months and leave the birds in peace to get on with raising chicks. It made for an extra busy year at Burwood, but by March 2018 the team was ready to move the first 18 birds north;

the remaining dozen would be moved a few weeks later, when they had finished moulting.

With the media watching on, the first tranche of Kahurangi-bound birds was gathered up from their enclosures at Burwood and put in individual travel cases. They were driven to Queenstown and then loaded aboard an Air New Zealand plane that had been specially chartered for the purpose. Each bird had its own seat and its travel box was secured with the seat belt during the flight north. Glen accompanied the birds and quipped, 'By the end, the entire cabin had the kind of special aroma that only 18 large herbivores in a confined space could produce!'[2] The plane landed at Nelson Airport, and while most of the birds were loaded into a helicopter to head directly into the hills, Lily and Rerehu, along with their 17-month-old helper Dalrachney – who may or may not have been their daughter (it's complicated) – were welcomed to the region with a pōwhiri at Onetahua Marae in Golden Bay. Here, Ngāi Tahu representatives handed over guardianship of their taonga to Manawhenua ki Mohua, the three Golden Bay iwi, before birds and guests were helicoptered to the release site.

The birds were released in various locations, each kitted out with a radio transmitter so they could be located over the following months and their health and progress monitored. The transmitters are designed to last five years before they need a battery change; to save battery power they operate for only 12 hours a day, during daylight. They emit 40 beeps per minute if the bird is moving, and after 12 hours of inactivity they change to 80 beats per minute, indicating that the bird is either dead or has lost its transmitter. Half of the takahē were also fitted with GPS loggers to collect detailed information about where they went over the next few months. To compensate for the extra weight, these birds had slightly smaller radio transmitters with a three-year battery life.

Along with the trio already mentioned, there were six other breeding pairs. Most, like Lily and Rerehu, were experienced parents who had reared a number of chicks and already passed on plenty of their genes to the population. Catlin had been puppet raised at Burwood from an egg collected in the Murchison Mountains in November 2005. Her mate, Shadowfax, had been collected from the Murchison Mountains in 2008 as a 10-day old chick and then hand-reared at Burwood. Together, they had hatched and raised 12 chicks, although two died within their first year. They had also produced two chicks during their recent stay at Burwood, but these didn't accompany their parents north – the male, Lowe, stayed at Burwood, while the female, Tuckett, was released into the Murchisons.

Other established pairs included Pipper and Matariki, who had produced nine chicks, including their son Scoop from the 2016 breeding season, who was one of a number of youngsters moved to Kahurangi. Scoop, of course, was named after the cadet reporter who broke the story of the rediscovery of the takahē back in 1948 (see Chapter 2). Pipper and Matariki, meanwhile, had a foster subadult from 2016, Tupuānuku, hanging out with them as helper. Hine Pou Pou and Nohorua had had five chicks in the time they had been mates, and their youngest daughter, Temple, was also on the move. There were also Uruao and Langley, with four chicks; Rusby and Tihaka, who had produced six chicks in their time together; and Hyde and Tametame, from Sanctuary Mountain Maungatautari and Mana Island, respectively, who had hatched in 2010, then met at Burwood and produced a single chick in 2014. The idea with these established pairs was that they were known breeders – if, for any reason, they didn't breed at Gouland Downs, it would be clear it was the site that was the problem.

In addition to the seven pairs, there were 16 single birds, some of which were youngsters aged one to two years that had yet to pair up and breed, while there were a few older, 'excess' females. All up, 19 females and 11 males were moved – and that sex imbalance may explain some of the shenanigans and restlessness that ensued.

Scoop exploring the Enchanted Forest on the Heaphy Track.
MADDIE VAN DE WETERING

The takahē team made regular trips to Kahurangi to monitor the birds and ensure they were settling in. They tracked them from the air and on foot, and were delighted early on to hear plenty of reports of sightings of Hyde and Tametame, whose new territory included a stretch of the Heaphy Track, where they were regularly seen by trampers. Lots of feeding sign showed that the birds were eating well, including the plentiful red tussock (*Chionochloa rubra* subsp. *occulta*) and plants such as *Gahnia*, which the birds had never encountered before. The takahē were so happy with their new diet that they showed little interest when the rangers enticingly shook jars of pellets, a noise that would have had the food-mad birds come running at Burwood.

It wasn't long before the transmitter signals showed that a reshuffling of partners was underway. Seven-year-old Te Uatorikiriki was an experienced mother of three who had been moved to Gouland Downs on her own, but she soon paired up with youngster Scoop, five years her junior. That relationship didn't last long, however, as Rerehu decided to 'divorce' long-time partner Lily, fight off Scoop and make a move on Te Uatorikiriki.

Then Rerehu also took a shine to one of the neighbours, Pipper, who upped and left her own long-time mate, Matariki, and moved in next door to form a threesome. She also took along her foster helper, Tupuānuku, bringing the happy family up to four.

Lily and Matariki, meanwhile, didn't waste too much time before they decided to get together. And in more swinging social news, young buck Scoop managed to lure Hyde away from Tametame, who eventually settled down with Temple near Saxon Hut – although these days Scoop is to be found near the Gouland Downs Hut with beau Erewhon. Phew! In the meantime, as spring approached, there were reports of fighting between some birds as they sparred over who got to live where.

The new threesome of Te Uatorikiriki, Pipper and Rerehu were all old hands at the breeding game. They got straight to it, and in mid-November 2018 their nest made national news headlines: a three-egg nest that was the first proof that the new Gouland Downs takahē population could be a success. But it would not be all plain sailing. The eggs were still at the incubation stage when takahē ranger Jason van de Wetering found Pipper's dead body during a routine monitoring trip. Unfortunately, the corpse was so decomposed that the cause of death couldn't be determined. Despite the loss, Te Uatorikiriki, Rerehu and helper Tupuānuku carried on incubating and went on to successfully raise a chick. This first Gouland Downs chick was a female, named Shiner.

This nest was one of three that first summer in Kahurangi, which coincided with a tussock mast and plentiful tussock seed to feast on. Langley and Uruao's chick was found dead a couple of days after hatching, but Hine Pou Pou and Nohorua had better luck. Their male chick thrived and was named Rocket in memory of the takahē detection dog that found it but that died several months later. From early on Rocket was a traveller – his parents moved him from Gouland Downs up into the neighbouring tops, before the trio

Shiner was the first takahē chick to hatch at Gouland Downs. Rocket the takahē dog that found her watches on. MARTIN GENET

were eventually caught by the takahē team and relocated back down to lower climes.

The takahē team were delighted that the first year in Kahurangi had gone so well – with only one death and two new chicks, the population had increased to 31. The most popular area for the takahē was proving to be the northern part of Gouland Downs, around Big River and Shiner Brook at the foot of the Slate Range.

The second breeding season at Gouland Downs got off to a great start. There were 10 pairs or breeding groups of takahē, and by early December 2019 there were seven nests. The male Rerehu was still with Te Uatorikiriki and former helper Tupuānuku, and their daughter, Shiner, hatched the previous year, stuck around to help as well. The foursome had a nest with three eggs, which suggested that both Te Uatorikiriki and Tupuānuku had laid, and at least one chick hatched.

Among the nesters were two other experienced pairs that had also hatched chicks the previous season: Uruao and Langley, and Hine Pou Pou and Nohorua. Then there were the first-timers, where one or both birds were new to the breeding game: Erewhon and Scoop, Temple and Tametame, and Muller and Tihaka. There was also a trio of youngsters all breeding for the first time: the female McKellar with partners Kauri and Taco.

Then the rain began to fall – lots and lots of rain, causing a flash flood through the low-lying area favoured by many birds. When the rangers next went in to check, they found that three nests had been washed out. The foursome lost their nest and chick, while the trio and Temple and Tametame lost their eggs. The remainder of the first-timers' nests all failed, probably due to inexperience.

The only glimmer of hope came from Uruao and Langley, who successfully hatched a chick. Sadly, it disappeared when it was about two months old.

After a thoroughly disappointing summer, a tracking flight in a light plane in mid-March picked up more unwelcome news. Most radio transmitters were sending out their usual cheerful 40 beats per minute, indicating the wearer was alive. But two birds were transmitting in mortality mode, at an alarming 80 beats per minute. A ground search a few days later revealed that 10-year-old Hyde had been killed by a predator on farmland outside the national park, and also discovered the partially decomposed body of nine-year-old Nohua, a kilometre away from her usual hangout near Gouland Downs Hut. At the time of her death, Nohua was on the move, having recently left her partner. Hers was one of a spate of divorces among former Burwood birds that were together as a result of 'forced pairings' made in the interest of improving takahē population genetics; now, given the choice, the birds were ending their relationships. Nohua was tucked under a hebe bush, and a nearby pile of droppings suggested she had been using it as a roost. Her body was sent for autopsy, but it wasn't possible to say why she had died. Although she was skinny, she had been eating right up to her death as her gut was still full of grass. The Gouland Downs takahē population now stood at 29 birds.

Out in the human world, the Covid-19 pandemic was just beginning to take hold and New Zealand was heading into lockdown. The plan had been to carry out an aerial drop of 1080 baits to knock down a growing plague of rats and stoats, but this had to be postponed. And as the global tally of Covid-19 infections and deaths began to climb, the Gouland Downs takahē population began its own bleak winter. In late May 2020, a post-lockdown monitoring trip revealed the corpse of nine-year-old Nohorua, high up on the tops. Like Nohua, he was tucked under a bush and was very light, although again he had been eating up until his death.

Then a midwinter monitoring trip discovered a sobering four further deaths. The older male Shadowfax, newly single, had wandered out of the trapped area and straight into the jaws of a predator. Rusby, aged 11, was freshly dead and very underweight – her death was attributed to emaciation. Eight-year-old Hine Pou Pou's underweight and badly decomposed body was found in a tangle of crown fern on the steep bank of a stream; it's hard to be sure, but it seems she may have got wedged and her death has been attributed to misadventure. Lily, aged 10, is also presumed dead through misadventure – her transmitter was in mortality mode and sending its signal from a steep gut that was too dangerous to check.

On 1 August, the Heaphy Track hut warden reported finding Tametame freshly dead, just a few days after the 11-year-old male was last seen alive and well. He was the fourth

The wanderers

While some of the Kahurangi takahē have been quite happy to settle more or less where they were released, other birds have turned out to have a wandering streak. Young Kapakapanui headed 14 kilometres west along the Heaphy Track to the James Mackay Hut, staying there for several months before making her way back to Gouland Downs. She then turned around and returned to James Mackay Hut, before walking back to Gouland Downs and then retracing her route back to the James Mackay Hut. She was still there in August 2020, when the 1080 drop took place, so she was taken back to the exclusion zone and kept in a pen for a few days to keep her safe.

Catlin, an older female, began moving around a lot after splitting with partner Shadowfax. She started walking east out of the national park towards the upper Aorere Valley, where her radio signal was picked up on a neighbouring farm during a regular aerial survey. Jason van der Wetering got in touch with the farmer, who took him out in his ute to help track her down so she could be returned to the safety of Gouland Downs. Catlin was put next to a former partner who was now single, but she showed no interest in her old beau. In winter 2020, the 14-year-old was on the move again, and by the time she was found in a bushy gully on a farm, well away from the safety net of traps, it was too late – she had been killed by an unknown predator. The killer had left tooth marks in her skull and neck, and the skeleton was picked clean. Her radio transmitter and cord harness were still wrapped around her spinal cord.

The news was equally grim for another single female with wanderlust. After breaking up with partner Tametame and following a brief dalliance with Scoop, nine-year-old Hyde had become highly nomadic. An aerial tracking flight in mid-March 2020 picked up a mortality signal from her transmitter, and her heavily scavenged body was found on the bank of the Aorere River, half a kilometre from Browns Hut and outside the national park. The Massey University vet who carried out the autopsy found a large puncture mark in her skull and speculated that perhaps Hyde had been killed by a ferret, the largest of the mustelids found in New Zealand.

Toothmarks in the back of the skull show that this takahē was killed by a large predator.
JASON VAN DE WETERING

takahē whose death was attributed to emaciation and poor condition. The relocated birds had all maintained their weight and stayed in good condition for the previous two years, so it was unclear why so many were now struggling.

A week later, the third takahē to be killed by a predator was found. Fourteen-year-old Catlin, like Hyde, was killed after walking out of the trapped area onto farmland (see box). It was a firm reminder that predators are bad news, and underlined the importance of the aerial 1080 operation. This was now rescheduled for mid-August, as part of DOC's nationwide Tiakina Ngā Manu – Battle for Our Birds response to the previous year's mega-mast event.

DOC was taking no chances. The 1080 bait drop would cover about 50,000 hectares of Kahurangi National Park around Gouland Downs and the Aorere Valley, but no bait would be dropped across the 587-hectare safe zone where most of the takahē lived. The plan was to gather up four birds – a pair and two singles – that lived in the drop zone and temporarily move them to safety. Kapakapanui, who had taken up permanent residence at James Mackay Hut, was duly collected and successfully held in a pen in the core takahē area, but the other three birds eluded capture and had to be left where they were.

The 1080 drop took place on 16 and 17 August 2020, with the takahē team closely monitoring 18 takahē to find out what happened. To the team's dismay, three-year-old Taco walked out of the safe zone during the operation and was found dead the next day, having eaten bait. Four days later, the transmitter of Maki, the single bird residing outside the core area, began giving a mortality signal – he, too, had eaten 1080 pellets.

Uneaten 1080 pellets persist on the ground for a few days until it rains, when the active ingredient is deactivated by water, so there was a window of risk for takahē. Unfortunately, during that risky period the young pair Erewhon and Scoop also temporarily wandered into the drop zone. On 23 August, Erewhon's body was found after Scoop was seen alone at Gouland Downs Hut – he had not eaten any bait, but his partner had, with fatal consequences. All up, six takahē were exposed to 1080 pellets – three of them ignored the pellets, and three died after eating them.

It was a difficult time for the takahē team. They had taken all reasonable precautions to keep the birds safe, short of not going ahead with the 1080 drop. But the loss of three adult birds to predators had also driven home why relying on the usual trapping regime was not enough in a mast year. The team already knew this from their experience in the Murchison Mountains, which also has a wide-scale trapping programme – when stoat numbers climb following mass beech and tussock seeding, 15–20 per cent of adult takahē are killed, and an even greater number of chicks and juveniles die from predation. DOC repeatedly says that evidence shows that 1080 is currently the most effective way of controlling rodent and stoat numbers at large landscape scale. The toxin will continue to be used as necessary in Kahurangi National Park, and the takahē team will try to keep the founding birds safe until, in the next decade or two, a new generation of takahē with no interest in pellets makes up the population.

The 12 deaths in 2020 brought the total number of deaths among the 30 founder birds to 13. In addition, the team had lost track of three birds, whose transmitter had either stopped working or been lost. Jason points out that it is almost impossible to find a takahē when it goes off grid, so we'll almost certainly never know their fate. There was some slightly better news: nine of the remaining 15 birds were a good weight and condition when checked at the end of winter.

All the losses meant that the third breeding season at Gouland Downs kicked off in late 2020 with just six breeding pairs or groups. The foursome of Te Uatorikiriki, Tupuānuku, Rerehu and helper Shiner were still together, as were Uruao and Langley. All the other pairs were newly formed: Temple with Scoop, Dalrachney with Kauri and Muller with Tihaka. There was also the surprise pairing of wanderer Kapakapanui and Moana – he was in the neighbourhood while she was temporarily in a pen for safekeeping, and they struck up a relationship despite the barrier between them.

There were only three nests. Dalrachney and Kauri – who resided near Saxon Hut – produced two chicks. Unfortunately, one of these died when it was month or so old, while the other one was almost the size of an adult when it disappeared after three months. The foursome nested and hatched at least one chick, and Temple and Scoop hatched a single chick, both of which survived for only about two months. It was another disappointing breeding season, in a year that was distinctly average for takahē across the country.

In March 2021, a new batch of 15 young birds, all one- and two-year-olds, was released at Gouland Downs to bring the population back to a more optimal 30. This is the greatest number of birds that can be easily monitored and managed here, yet still give them ample opportunity to find a mate of their choosing. It will be a year or two before the new arrivals finally settle in, pair off and begin to successfully breed. And it will be difficult to monitor this success, as six of the 15 promptly went off the radar despite being fitted with radio transmitters. Jason searched a large area from a fixed-wing plane in case the birds had wandered far afield, but he suspects they either lost their transmitters or these stopped working. Two of the new birds are known to have died.

Dalrachney, who arrived as a youngster with the first cohort of birds, has proved rather unlucky in love. Her first mate, Shadowfax, died from predation. Her next partner, Kauri, was found emaciated and dead in July 2021. She then partnered up with Fulton, one of the latest young arrivals, but the pair didn't breed and the relationship faltered after Fulton got itchy feet and started moving around.

Overall, the fourth takahē breeding season at Gouland Downs was another lacklustre affair. It started on a back foot with the loss in late winter of its experienced breeding trio: the two females Te Uatorikiriki and Tupuānuku, and the male Rerehu, who had successfully bred in the first breeding season. The tight-knit group all drowned in a flood, with signals from their radio transmitters leading ranger Martin Genet to their bodies in Big River.

That left just four breeding pairs. For the second year in a row, Uruao and Langley and Muller and Tihaka failed to nest. Now in their second year together, Temple and Scoop got off to a promising start in December, nesting and laying one egg. But by the end of January there was no sign of an egg or chick, suggesting that either a weka had eaten the egg or that the chick had died when it was very small. A new couple, McKellar and Moana, initially had more success, hatching a single chick. However, that chick disappeared when it was about two months old. There were several major rain events and serious flooding in Northwest Nelson in early 2022, which damaged or washed away three bridges at the western end of the Heaphy Track. Luckily, the takahē population came through these adverse weather events unscathed.

The jury is still out as to whether a self-sustaining population of 30–50 takahē will develop at Gouland Downs. At the time of writing, the only successful breeding season to date was the first one, which coincided with a tussock masting year. That, and the deaths of a number of emaciated birds in poor condition, really does beg the question of whether the low-fertility 'hungry country', as the area is sometimes called, grows plants of sufficient

One of the first 30 takahē to be released in Kahurangi National Park in 2018. JAKE OSBORNE

quality to support breeding. It will be interesting to see if breeding is more successful in the next tussock mast year – whenever that is. And if it does turn out that mast seeding is crucial to takahē breeding in Kahurangi, then that may not result in enough chick production to counteract deaths and allow the population to be self-supporting in the long term.

For now, the takahē team are taking a 'wait and see' approach. They don't want to call it quits too soon, and besides, where would the Gouland Downs birds go, at a time when there are plenty of other takahē also needing homes? And if all does go well, with the Gouland Downs population really taking off and the issue of 1080 use for predator control resolved, then down the track the takahē team could look to expand the population to neighbouring Mackay Downs and Gunner Downs in Kahurangi National Park.

In the meantime, the Gouland Downs takahē have been a tremendous advocacy success with the thousands of trampers and mountain bikers who use the Heaphy Track each year. People thoroughly enjoy the opportunity of coming across a wild takahē, often near one of the huts where the grass is short and lush. Some locals from Golden Bay have been visiting regularly just to keep up with the new residents. It turns out that takahē are a 'great attraction' on a Great Walk.

CHAPTER THIRTY-ONE

OROKONUI

Orokonui Ecosanctuary, near Dunedin, is one of the smaller takahē sites. It is mostly covered in forest, and the only places for takahē to establish a territory are the upper lawn near the visitor centre and another area at the lower end of the sanctuary, made up of a wetland and a grassy area that runs along the predator-proof fence. It's boutique, but as long as takahē pairs living in these two territories are fecund, producing chicks that can be sent elsewhere, then that is enough for the ecosanctuary to be part of the takahē metapopulation. The bonus of Orokonui as a site is that the upper takahē territory is where most visitors head to, so there is plenty of opportunity for people to encounter a takahē or two.

The first takahē to live at Orokonui arrived in May 2012. They were a couple of middle-aged males, fondly – and not inaccurately – referred to as 'duds'. For Quammen, named after American nature writer David Quammen, it was not for want of trying. The 10-year-old male hailed from Maud Island, and he and his former mate had produced 18 eggs, not one of which had hatched. Quammen arrived with fellow Maud Island local Te Hoiere, who was a year younger. As a youngster, Te Hoiere had partnered with a slightly older female, Hilda, and in 2004 they had produced a male chick, Hauhanga. But after that promising start, Te Hoiere had failed to produce any more offspring, so he was packed off to Orokonui to free up his Maud Island territory and to give him a fresh start.

Te Hoiere was bossy and didn't survive long before succumbing to septicaemia, but Quammen was not left lonely as he soon paired up with the sanctuary's third arrival. This was a distant female cousin called Paku, whose name means 'small'; she was hatched on Kapiti Island in 2004 and raised by her parents. In 2007, Paku and her mate Bellamy raised a female chick named Kawa, who would go on to be well known as the slightly overweight, sweetcorn-loving mate of infertile Tumbles at the Te Anau Wildlife Park, where they were diligent foster parents.

Members of large and genetically common families, Quammen and Paku were not

Takahē chick being fed by its parent. CRAIG MCKENZIE

expected to produce any offspring; their role was entirely ambassadorial. But that didn't stop them from trying. Paku laid repeat infertile clutches in 2013 and again in 2014, and the pair were diligent with incubation, remaining on duty long after the due date. By 2015, the takahē team decided to take advantage of their dedication, and their infertile egg was substituted with a fertile egg from Burwood. Everyone was delighted when a chick, Kotahi, hatched; local schoolchildren suggested the name, meaning 'number one', for the sanctuary's first takahē chick, who was initially thought to be female but later was confirmed as a male. Quammen and Paku excelled in their chick-rearing duties, and Kotahi was left with his foster family for the winter, the idea being that he would remain as a helper for the next lot of foster eggs.

When Paku and Quammen nested in 2016, it was assumed that the resulting two eggs would be infertile as usual, and plans were made to bring two fertile eggs from Burwood for them to raise instead. But when the day came for the substitution of fertile for infertile eggs, there was a surprise – one of Paku and Quammen's eggs was fertile! It was much less developed than the two Burwood eggs, and so couldn't be left as part of the same clutch. The short-term solution was to foster it out to a young pair of takahē, Marokopa and Mohio, who were now occupying the lower territory at Orokonui. But there were concerns about this pair's lack of experience and the egg was moved again, this time to the Te Anau Wildlife Park, where it was entrusted to foster parents Tumbles and Kawa. Being Paku's daughter, Kawa was now set to raise her half-sister. Orokonui staff and supporters dubbed this surprise chick their 'miracle' chick, and when she hatched, she was named Ehara, an expression of excitement meaning 'great'.

Back at Orokonui, Paku and Quammen successfully hatched their two foster eggs. About this time, they decided to dispense with the services of helper Kotahi, who they chased away. Now surplus to requirements, Kotahi was sent to Burwood, and the doting foster parents got on with raising the female Tumanako, whose name means 'hope' or 'wish', and the male Wheko, meaning 'chirping of birds'.

But Paku and Quammen weren't finished with their surprises. The following year they succeeded in producing another fertile egg, and they were left to raise this one – a male named Mihiwaka – themselves, with some teenage help from daughter Tumanako. Son Wheko had been relocated to Burwood, where he was already paired up with a female called Jenkins.

In 2019, the Orokonui takahē were not so lucky. Paku and Quammen were caring for two young foster chicks when heavy rain began to fall, causing significant flooding in the sanctuary. By the time the rain stopped, the nest was empty and both chicks had succumbed to the wet and cold. The verdict: bad weather, bad luck and perhaps a spot of bad parenting.

By 2020, Paku and Quammen were considered past their breeding prime and were retired to a quiet life in the lower territory at Orokonui. Stepping into the spotlight in the upper territory were Burwood residents Bennett and Waimarie. Bennett was named after Ted Bennett, a young Southlander who spent a lot of time on expeditions in Fiordland with Geoffrey Orbell and who was with him in Takahē Valley just two weeks after takahē were rediscovered there. A neat link is that Ted is also father to an employee of Fulton Hogan, the Takahē Recovery Programme's corporate partner.

Originally, Orokonui had been intended just as a display site for birds that were either too old or infertile. But it had shown that it was a good place to raise takahē chicks, and the latest thinking by the takahē team was that even small sites could play an important breeding role. The new arrivals, therefore, were just three and two years old, with a potentially long breeding career ahead of them. The two spent no time in getting down to the business of making babies, and in spring the same year they incubated their first eggs and successfully hatched a chick. Unfortunately, the chick was born with a defect and did not survive.

In early spring of 2021 the pair were nesting again, and by mid-October the two chicks had already left the nest and were wandering around the sanctuary with their doting parents, learning how to eat natural foods and how to turn up punctually for the 11 am daily public feeding. At the time of writing, one chick has been named Kapua, while the other one awaits a name, and both have been sent to takahē school at Burwood. Their future has not yet been decided, but it's clear that their parents, Waimarie and Bennett, will continue at Orokonui for the foreseeable future, fulfilling a dual role of takahē ambassadors and makers of baby takahē. They are so enthusiastic that in August 2022 they were found nesting, the earliest takahē on record to start breeding.

It's interesting to see where Paku and Quammen's chicks and foster chicks have ended up in the wider takahē diaspora. Kotahi spent a few years at Burwood, where he was partnered with Weydon, who was nine years his senior and had had two former partners. She has produced 17 chicks, five of which were fathered by Kotahi. He was just two years old when he had his first chick, so it was lucky for him he had such an experienced partner. By the by, Kotahi's situation mirrors that of his biological father, Charles, who was much younger than his mother, Widget. In 2021, Kotahi was moved to Wairakei, to live on the golf course. His new partner is George; she is four years his senior and, including the 2021 chick she produced with Kotahi, has had 15 offspring.

Paku and Quammen's 2016 foster daughter Tumanako is at Burwood, and she and partner Bendigo are yet to produce any chicks. Her brother Wheko, on the other hand, is a precocious and prodigious breeder. At age two he fathered his first chick with Jenkins, followed by a further two a year later, a singleton the following year and then another pair – not a bad tally for a five-year-old! He can probably be excused, as he has been aided and abetted by the takahē team – they are keen to maximise his rare genes, passed to him courtesy of biological parents, Catlin and Shadowfax. Wheko didn't breed in the summer of 2021/22, when he was no longer paired with Jenkins.

Ehara, the 'miracle' chick, is considered of low genetic value because of Quammen and Paku's common genetic lineages, and she is one of the many young birds that have been transferred to Gouland Downs, where she remains single. Her younger brother Mihiwaka, who hatched in 2017, was sent to Burwood for a winter of tussock and *Hypolepis* training before being dispatched to the Murchison Mountains, where he lives a reasonably incognito life.

The comings and goings of the Orokonui birds is a microcosm of what has been happening with the wider takahē population. The Takahē Recovery team have worked to maximise genetic diversity, spread the risk by having birds at multiple sites and create new opportunities for people to experience one of New Zealand's rare endemic birds. But not every venture has worked as well as this one.

CHAPTER THIRTY-TWO

DISAPPOINTMENTS

Tāwharanui Open Sanctuary, north of Auckland, has had so many takahē breakouts that they've had to institute a 'two strikes and you're sent to a real island' policy. This is a shame, as it's a popular place with lots of visitors from Auckland coming for the sandy beaches, the warm sea and the opportunity to see rare wildlife such as the North Island saddleback and takahē.

Tāwharanui Open Sanctuary is part of Tāwharanui Regional Park, and is a peninsula that has been cleared of predators and is protected by a predator-proof fence. But the fence has a weakness that some takahē have been quick to exploit. Instead of being a complete enclosure, the fence runs in a straight line across the peninsula, terminating at a sandy beach on each side. To try to prevent predators such as stoats and cats walking into the sanctuary past these 'leaky' ends, the fence spirals outwards at each extremity, the idea being that an animal walking along the fence line finds itself spiralling into a dead end filled with an array of traps all baited with enticing lures. This design works well to prevent creatures on the outside of the fence reaching the inside, but it does nothing to stop a takahē intent on walking out – which is what a surprising number of the birds have done.

The Auckland Council rangers who manage the sanctuary usually know now when one (or two) of their takahē residents are leaving, because an electronic sensor on the fence detects the birds' radio transmitters and sends a signal. Outside the sanctuary, the birds find themselves navigating farmland, backyards and vineyards, all with plenty of pasture grass to feed on. Sometimes it's a simple matter for the rangers to gather up the wanderers, but at other times it can take a few days to track them down and then use food to lure them into cages.

The returnees are generally kept in a pen for a few days in the hope their wanderlust wears off, but this doesn't work for some of them. Once they've had a taste of freedom, they often make a second or even third attempt to escape. Unfortunately, the big, wide world

Kiko, Frana and Tamahunga in kikuyu grass in Tawharanui Regional Park. ALISON STANES

beyond the fence is full of dangers such as cats and dogs and cars, so after two escapes the birds are rehomed to Motutapu Island in the Hauraki Gulf / Tīkapa Moana, a real island surrounded by a wide watery moat.

Well-known Tāwharanui escape artists include Peni, Pleiades, and a pair called Walter and Manaaki. The adventures of the latter took them more than 20 kilometres, and their escapades outside the fence inspired the 2021 children's book *Takahē Trouble!*, written by Sally Sutton and illustrated by Jenny Cooper.

The first takahē arrived at Tāwharanui in 2010, and the takahē team was hoping that the site would be able to support a moderate-sized breeding population. Unfortunately, the practical realities of the leaky fence and the number of escapes, as well as concerns about potential interactions between takahē and the high numbers of visitors' cars, stress from frequent summer droughts and diseases such as erysipelas, mean that the sanctuary has not met expectations. While some birds will remain, Tāwharanui will no longer rank as a key takahē site.

Another northern site, Motutapu Island in the Hauraki Gulf / Tīkapa Moana, has also been a disappointment in terms of breeding. Takahē were first introduced to 1,510-hectare

Mistaken identity and tragic consequences

While the pūkeko is quite different to the takahē in many ways, people who are unfamiliar with birds can easily confuse the two. This confusion had its most tragic consequences in 2015, when four takahē on Motutapu Island – Raumati, Ariki, Westy and their chick (the island's first) – were shot. The case of mistaken identity was made by hunters from the New Zealand Deerstalkers Association who were helping DOC with a pūkeko cull. Pūkeko on the island number in their thousands and are a nuisance as they pull out large numbers of seedlings being planted in an effort to restore native bush following a history of farming on Motutapu. The culls have been a regular event to maintain the pūkeko population at a manageable level. The hunters mistook the large flightless takahē for the much smaller pūkeko, despite a detailed briefing about the two species and being told to shoot only birds that were in flight.

Motutapu in 2011, when it and adjoining Rangitoto Island were declared pest-free following a 20-year eradication effort that saw the removal of nine invasive pest species. Motutapu used to be farmed and has extensive areas of pasture grass, and the takahē team had hoped the island would be a net producer of birds and able to support about 20 breeding pairs. However, it is not living up to that promise.

The loss of a takahē breeding group, including a chick, on Motutapu, when they were mistakenly shot during a pūkeko cull in 2015 (see box), was a significant setback, but even without that there have been more deaths in the population than there have been births. Then there was the discovery of stoats on Motutapu and Rangitoto in May 2020. The invaders evaded capture for a quite a while, but over the following 18 months three stoats were eventually caught in traps. All were male, and it is thought that they swam over from the mainland or nearby Browns Island (Motukorea). While a stoat did kill several shore plovers, there is no evidence that any takahē were victims, but the presence of the mustelids was still a worry.

Like Tāwharanui, Motutapu will remain a takahē home for now, but only for a small number of birds that are considered surplus to the main takahē programme – in other words, those with the most common genes. Part of the problem for the two sites' poor performance from a breeding point of view may be that they are simply too far north and too warm for takahē, which of course hail from the southern South Island.

Tāwharanui and Motutapu, along with Auckland Zoo, might not be considered valuable sites for takahē breeding, but they will continue to play an important advocacy role as they are on Auckland's doorstep and provide easy opportunities for people to see takahē.

And other islands are actually doing well. Mana Island, which now holds just eight takahē pairs, reliably produces good numbers of chicks, as does a Ngāi Tahu-owned island in Foveaux Strait.

Takahē are popular residents in the busy Tāwharanui open sanctuary. **ALISON STANES**

CHAPTER THIRTY-THREE

THINK BIG

By 2019, takahē numbers were on a roll. Just three years after it exceeded 300, the population passed 400 birds. The annual rate of increase varied, but it was hitting some new highs: after a very satisfying 10 per cent in 2016, it reached a dizzying 13.4 per cent in 2017. Yet another Murchison Mountains stoat plague in 2018 dropped the overall growth rate, but it was still a healthy 8.4 per cent. Most importantly, the number of breeding pairs was growing quickly as young birds reached breeding age. In 2018, the total number of breeding pairs rose to 112 as a number of young birds came of age. The following year that number climbed to 130 breeding pairs.

Burwood was key to this success. In 2018, for example, the 19 breeding pairs in residence churned out 32 chicks, 24 of which reached subadulthood and were counted as new members of the takahē population come 1 October 2019. In total, 54 chicks were produced across monitored sites, which was the first time the team had cracked 50 chicks in one year. And it wasn't all about quantity – genetic quality was also being significantly improved at the same time.

But – and there always seems to be a 'but' – the mega-mast was well underway and the resulting stoat plague hit the Murchison Mountains and Gouland Downs in the 2020/21 breeding season. We've already heard of the effects of the mega-mast on the Kahurangi population, and as expected, it had a devastating impact on the Murchison Mountains population as well. Even with full-scale stoat trapping, the takahē team now expects to lose 10–20 per cent of the wild birds in a bad stoat year – which at least is much better than the devastating outcome in 2007.

Combined with the poor 2020/21 breeding season were the unexplained deaths of a number of island birds. But despite all this, numbers at the end of the year were just slightly down, with 440 birds rather than the 445 there were at the beginning. As Glen

Greaves from the Takahē Recovery team says, instead of 'periodic plummets' there are now periodic plateaus.[1]

To help offset the losses (and because there was nowhere else for them to go), all the surplus young birds being produced at Burwood and at the other productive sanctuary sites were being poured into the Murchison Mountains, with a few going to Gouland Downs to keep that population at 30. In non-stoat years the Murchison Mountains birds are raising good numbers of chicks themselves, and as a result, the area is bulging at the seams – although Glen says the team now believe that the Murchisons can hold more birds than they originally thought. However, they are still guessing at the carrying capacity, and they don't want excess birds spilling out to the west, outside the trapped area. They also don't want too many birds overgrazing some areas and then having to move on to find new, greener pastures. By mid-2022, it was estimated there were 218 birds in the Murchisons. The next time the takahē team head into the area to carry out one of their three-yearly ground counts will be 2023; in the meantime, they continue to monitor a sample of about 40 radio-tagged birds using Sky Ranger.

Unfortunately, at the time of writing DOC has been grappling with funding issues as an ongoing consequence of the Covid-19 pandemic and has been looking for ways to save money. One of the things on the potential chopping block was regular stoat trapping in the western periphery of the Takahē Special Area. There are only 16 or so birds living here and the team has had difficulty justifying the ongoing expense of stoat trapping there, even though they suspect it's a useful buffer zone against stoats invading from further west. In addition, stoat trapping offers no benefit to smaller threatened species such as bats and rock wrens, for which rat trapping would be more useful. The trapping programme has had a reprieve for now, but it will remain a candidate for possible future cost saving.

Another potential casualty of funding cuts was deer control. For the past few years commercial venison hunters have been subsidised by DOC to hunt deer in the Murchisons, but a crash in international venison prices meant it was no longer financially viable for them to fly in and there was no DOC money available for the job, although some aerial deer control was carried out in a buffer zone to the west of the Murchisons. However, a recent study has highlighted how controlling browsing mammals such as goats and deer to low levels in native forest not only helps save biodiversity values, but also helps the vegetation sequester large amounts of carbon dioxide, which would help in New Zealand's effort to tackle climate change. This has led to a renewed interest in deer control, which was reflected in a tranche of money being made available for this activity in the 2022 Budget. At the time of writing, it is not clear what this money means for deer control in the Murchisons. Personally, I think stopping deer control in the Murchisons would be short-sighted. As discussed earlier in the book, high deer numbers have a devastating impact on plant cover and the subalpine vegetation is slow to recover from heavy browsing. And it shouldn't be forgotten that the importance of high-quality habitat was one of the significant findings of the intense research programme in the 1970s.

The future of pest control in the Murchisons will ultimately be decided in terms of what's best for the whole ecosystem, which is home to threatened kea, whio, Fiordland tokoeka, bats and rock wrens, as well as takahē. Because of the presence of takahē, 1080 has never been used here, but there are strong arguments for its use as it controls rats as well as killing stoats through secondary poisoning. Stoats are patchily distributed, so the idea would be to target the forested areas with the highest numbers. Alpine areas and key takahē country would need to be excluded, at least until the problem of takahē eating pellets has been solved. Further pellet trials with captive pellet-naive takahē have suggested that birds that have never been fed pelleted supplementary food simply ignore the cereal baits that 1080 is mixed with, but the sample sizes are still too small to draw firm conclusions. Unfortunately, it isn't as simple as stopping supplementary food. At many sites – especially Burwood, where the birds are constrained within pens and can't roam widely, and where the copper tussock isn't as nutritious as other *Chionochloa* species – the birds rely on this extra food to get them in breeding condition.

To help solve the problem, nutritionists at Massey University have been asked to develop a new takahē food that is not based on grain, and has a different colour and scent to cereal-based pellets. It's particularly important that it doesn't come in pellet form, and a mash isn't suitable either as 1080 pellets turn to mush when they get wet. The researchers are looking at an existing gluten-free horse food that could be tweaked to perfect it for takahē, and are trying to work out if this could be delivered as a flake – think large bran flakes. There is also new research looking at delivering 1080 in a meat sausage instead of a cereal pellet. This would be a more targeted way of killing only rats and stoats.

Front of mind for the takahē team recently has been where birds can go next, as all the existing sites are full. The takahē metapopulation encompasses 20 sites at the moment, but that includes small retirement spots such as Auckland Zoo and productive mid-sized sanctuaries such as Mana Island, as well as the Murchison Mountains and Kahurangi National Park. Maintaining the genetic quality of birds and relationships with people at all those smaller sites takes a lot of work, so the team are thinking big and are urgently looking for large areas with existing predator control.

Fiordland's Stuart Mountains have been ruled out as there is no money to roll out a new trapping network there. The Greenstone Valley will almost certainly be first off the rank. A popular walking track follows this wide, gentle valley at the head of Lake Wakatipu in southwest Otago, and is often linked with the Caples and Routeburn Tracks as circuits. It neighbours Fiordland and Mount Aspiring National Parks, but the land is a working farm belonging to Ngāi Tahu, who already partner with DOC on the Takahē Recovery

Greenstone Valley. DANILO HEGG

Programme. From a takahē point of view, there is a rich mix of tussocks and European grasses. And from a takahē ranger perspective, the area is quite close to the operations base in Te Anau and there are already good tracks and huts. There is also an existing trap network, which is growing larger. Trapping carried out so far shows lower stoat numbers than in the Murchisons, but on the negative side of the ledger there are feral cats. It's not known how takahē will be affected by cats, and increased effort is being put in to remove them.

The Greenstone Valley sounds a very suitable location, but despite its large size it is estimated that it might be able to hold just 30 pairs. At the current rate, that will account for only the next two years of takahē production.

After that, the takahē team is eyeing the much larger Rees Valley, slightly north of the Greenstone. The Rees and Dart / Te Awa Whakatipu valleys form another popular tramping circuit, and while part of the track lies in Mount Aspiring National Park, the lower reaches of the Rees are privately owned by a few farms. The landowners have already begun trapping and habitat restoration, and have reacted positively during initial discussions about takahē. In time, there is the possibility that the Greenstone and Rees populations could be linked in some way.

Burwood will play a key role in training northern-born birds for release into the Greenstone and Rees Valleys, teaching them to eat tussock and getting them accustomed to cooler southern weather. This means it will continue to be something of a bottleneck in terms of the transition of takahē from northern sites, as it can only hold so many birds.

Sanctuary sites that act as insurance populations continue to be a key focus for the takahē programme, even though managing small numbers of birds at multiple sites takes a lot of effort. All the common and inbred genetic lines of takahē have been cleared off islands in a series of big moves, and the idea now is that the key islands and mainland sanctuary sites will collectively be home to 65 breeding pairs of takahē that are the genetic cream of the crop. But where are all those common northerners going to go? What is needed is a large northern site where all the takahē that don't have an important genetic role to play and that are too old to retrain for release down south can live out their days. In the interim, Motutapu can take quite a few birds, and if it is essentially a takahē retirement home it won't matter if breeding there continues to be poor.

One possible site is Capital Kiwi, in Wellington. This is a large area of coastal scrubland on the southwest corner of the North Island, partly owned by Wellington City Council and partly farmed. The large network of stoat traps in place has reduced numbers of the predators to very low levels, and kiwi have just been released there. The habitat looks suitable for takahē, and if the kiwi release is successful, then it would make an excellent takahē destination.

What takahē really need, though, is a predator-free New Zealand. In 2016, the National government launched Predator Free 2050. Its ambitious goal is to rid New Zealand of all rats, possums and stoats by 2050. This means the complete eradication of Norway rats, ship rats and Pacific rats, as well as possums and three species of mustelids – stoats, ferrets and weasels. The effort encompasses everything from backyard trapping groups, to ambitious projects such as Predator Free Rakiura, to companies such as Zero Invasive Predators (ZIP) working at a landscape scale to see if they can develop techniques to eliminate predators and then keep them out of large areas of backcountry. A lot of research is being carried out to find more efficient ways of luring animals into traps, and to develop and deliver toxins that target specific species. Needless to say, there are plenty of threatened species recovery programmes that would like to see Predator Free 2050 become a reality, and the takahē team is having frequent conversations with people working in this field.

One possibility that will hopefully be available for use soon is a new toxin known as PAPP. PAPP is short for para-aminopropiophenone, and is a vertebrate pesticide that was developed in New Zealand to control carnivores. It was licensed in 2011 and approved in

2014 for the control of stoats and feral cats; it's also being looked at for ferrets. Overseas, PAPP is being considered for coyotes, foxes and wild dogs. After eating the poison, animals die within a couple of hours from a lack of oxygen to the heart and brain. Although PAPP is toxic to takahē, they would have to eat three to four baits for it to be fatal and since it will be delivered at a bait station encapsulated in a meat sausage at a rate of about one bait every 20 hectares, it's highly improbable the vegetarian birds would eat a toxic dose. Like 1080, PAPP is water-soluble and breaks down quickly. A device known as the Spitfire has also been under development for a few years: when a stoat runs through a tunnel, which they like doing, PAPP is sprayed on their belly and they then consume a lethal dose of the toxin when they groom themselves. It's a long and tricky process working out how to safely deliver a new toxin, but hopefully it won't be much longer before it can be used.

Glen says it's important that future management of takahē is not reliant on technology that hasn't been developed, so the takahē team are not hanging hopes on unproven – and controversial – techniques such as gene editing. But at the same time they aren't afraid of taking risks, in the hope that some will pay off, and they know that useful lessons can come from failure. Glen and the team also understand that some things will take time and patience, such as finding out whether Gouland Downs offers good long-term prospects for its takahē.

It's taken 75 years for the Takahē Recovery Programme to get this taonga to where it is today – tantalisingly close to 500 birds that are in great genetic shape and with a good rate of annual population growth. The takahē is a rare bird that's also increasingly available for people to see and enjoy, which can't be said for many of New Zealand's iconic threatened species. Like other long-term conservation programmes, such as that supporting the kākāpō, the Takahē Recovery Programme has been a veritable rollercoaster ride for the species and the people caring for it. There have been plenty of dashed hopes and challenges along the way, and at times progress has seemed slow. Overall, however, the programme has been – and continues to be – an inspiring success story that deserves to be better known.

By the time the rediscovery centenary comes around in 2048, takahē will hopefully number in their thousands, will be inhabiting large areas of predator-free grassland, and will no longer be considered at risk and consequently only lightly managed. And I like to imagine a more distant future in which takahē still roam the magnificent wilderness of the Murchison Mountains, mast years are times of plenty and stoat plagues are but a dim memory. In this takahē utopia, the birds are also spread across the country, with free-ranging wild takahē grazing local community parks and gardens, and adding their shrill duets to the growing richness of the dawn and dusk bird chorus. And small children dream about the giant blue birds and the remarkable story of how these taonga were rediscovered and rescued from extinction.

CHAPTER THIRTY-FOUR

WHEN DREAMS COME TO LIFE

The wide, flat floor of Takahē Valley has been scoured by many glaciers. As I walk up the valley, I try to imagine it in the past, with hundreds of metres of ice above me. Today, it is mostly a wetland, a mosaic of small tussocks riding atop hummocks, the curvaceous mounds of cushion plants and pillows of apricot-coloured Sphagnum *moss interrupted by sparse branches of low hebes. The ground is starred with tiny yellow flowers, the first of the spring alpine blossoms. There are sundews unfurling red fronds covered in hairs tipped with sticky drops to snag passing insects. Small streams thread sinuously across the floor, like veins generously watering the valley. As I carefully make my way between the slightly firmer patches of ground, I lurch forward suddenly as one leg disappears up to the knee in a soggy hole, created when the base of the last glacier to pass this way plucked out a boulder as it ground relentlessly overhead. The steep sides of the valley, the rounded head basin and a few cirques or hanging valleys perched high above are further reminders of the sculpting power of frozen water.*

Tawny golden-green hebes and chestnut-brown Dracophyllum *stripe the ground alongside me. Bands of sombre-green beech forest line the lower valley sides, and ahead I can see long scarves of dark grey scree draping down the cliffs. On the ridges there is a spattering of snow left behind by last week's late storm, but today the sky is clear and the spring sun is warm.*

Takahē have walked here before me. Earlier, I had marvelled at splayed takahē toeprints in soft sand on the shore of Lake Orbell. Now I can see discarded tussock stems, dropped after the bird has snipped off the tender base. There are plenty of droppings, too, like miniature hay bales. I spot tunnels under the thick tussock and imagine a takahē taking a secret shortcut before striding out, stepping lightly across cushions of moss that I would plunge straight into. I scan the valley ahead, looking and listening to see if I can locate any takahē, but I hear only a gentle wind, a distant pipit, the suck and splash of my feet and my own slightly laboured breathing.

Off to my right, I can see Glen Greaves and his dog Yuki taking a parallel route to mine. We regularly stop to play takahē calls, and a few hundred metres past the head of Lake Orbell our recorded call solicits a prompt reply. I wait while Glen and Yuki edge closer. They stop moving and then stand quietly, watching something in the low vegetation.

Glen waves me over. 'It's a very friendly family,' he says. 'One adult doesn't have a band and there is also an unbanded juvenile that came out to inspect Yuki.' Unbanded means a bird that has never been handled by a human. Birds that have been caught and banded tend to be much more wary of humans and run away, but it seems that these ones – especially the youngster – are happy to stick around rather than flee. I move forward very slowly, trying not to make any noise. There is plenty of feeding sign and a very fresh green dropping. Then I see a movement to my left, something blue behind an obscuring veil of tussock. I crouch as the blue materialises into a young takahē that steps forward towards me, curious but wary. We're probably the first humans (and dog) this young bird has ever seen. We watch each other for a few long seconds, and I'm reminded of Geoffrey Orbell and the rediscovery team, and the first takahē they encountered in this valley. What a memorable moment that was for them – a bird and a person, each getting their first glimpse of the other.

I can hear the juvenile's parents making contact calls, and it soon takes heed, turning to disappear into a thicket of hebes. My grin stretches across my face as I turn to Glen. 'That was an Orbell moment,' I say, and he nods in agreement. Nothing beats the thrill of meeting a wild takahē on its own terms in such a stunning environment. While few of us will have the opportunity of visiting the Murchison Mountains, I hope that in a few years, as wild takahē populations become established at new and accessible sites, more people will have their own Orbell moment when they meet the bird of dreams in a remote wilderness.

ACKNOWLEDGEMENTS

It has been an immense privilege as a writer to tell the stories of Aotearoa's two surviving flightless giant birds: takahē and kākāpō (told in my earlier book *Kākāpō: rescued from the brink of extinction.*) Both birds were in a downward spiral that could easily have ended in extinction, had not talented people worked hard over many decades to ensure not just their survival, but a secure future. Many of those people have contributed to this book, and I am deeply grateful to them for sharing their knowledge, their experiences, their wisdom, and their photos. Seventy-five years is a very long time for an active conservation programme to be running and the full encyclopaedic detail would have run to many volumes, so I have only been able to share some of this wealth of information in a single book. All omissions and any unintended errors are regretted and are entirely my responsibility.

Deidre Vercoe is Manager of the Kākāpō and Takahē Recovery Programmes at the Department of Conservation, and she deserves special mention for encouraging me to write this book and for providing much appreciated support in many ways. Senior Ranger Glen Greaves also merits particular recognition for providing opportunities for me to join the takahē team in the field, and for patiently providing documents and answering endless questions over a number of years. A huge thank you to the rest of DOC's Te Anau-based takahē team, for cheerfully sharing so much with me: Nichollette Brown, Phil Marsh, Jason van de Wetering and Julie Harvey. Thank you, too, to team members at the Burwood Takahē Centre: Lisa van Beek, James Bohan and Sam Haultain.

I am grateful to many people who have worked in takahē conservation for sharing some of their stories: Jane Maxwell, Daryl Eason, Andrew Digby, Kerri-Anne Edge, Nick Torr, Dave Crouchley, Peter Moore, Andy Garrick, Jim Mills and Roger Lavers. Having started work with takahē in the 1960s, Brian Reid wins the prize for having the earliest direct experiences with 'tarks'.

I would also like to acknowledge all the other dedicated people who played pivotal roles in the takahē team over the years – I may not have spoken with you directly, but your legacy of reports and papers has been invaluable.

I benefited from the knowledge of a number of scientists who have been involved with the takahē programme: Alan Mark, Bill Lee, the late Ian Jamieson, Lara Urban and Annie West.

Thanks to Mana Island rangers Nick Fisentzidis and Genevieve Spargo. And a shout-out to volunteers and staff at sanctuaries around the country for recording the exploits of 'your' takahē, which I have liberally borrowed.

To all the photographers who willingly shared so many interesting images, thank you for your skills and, in many cases, for lugging cameras around in already full backpacks in difficult Fiordland terrain and weather. Rod Morris and Ross Curtis deserve special mention. Thanks, too, to the image librarians at DOC in Wellington and at Archives New Zealand, the Alexander Turnbull Library and other archives around the country for keeping and caring for important historic photos.

It was a pleasure to meet Joan Watson, the last surviving member of the 1948 rediscovery party, and a privilege to join family members of that party when they released a pair of takahē in the Murchison Mountains to mark the 70th anniversary of the rediscovery.

The production of this book has benefited greatly from the team at Potton and Burton. Susi Bailey did a superb job of editing the text. And I owe a debt of gratitude to publisher Robbie Burton, not only for his skill and expertise in bringing the book to fruition, but for his patience and understanding when personal events derailed promised deadlines.

To my partner Malcolm Francis, thank you for your love and support during what has been a challenging few years. The companionship and support of friends has been invaluable, especially Ash McCrone, Mandy Armstrong, Lianne Parkin and Laurence Fearnley.

Takahē head. JAKE OSBORNE

ENDNOTES

PREFACE

1 John W. Fitzpatrick, foreword to *The Takahe: Fifty Years of Conservation Research*, eds William Lee and Ian Jamieson (Dunedin: University of Otago Press, 2001), pp. 9–10.

CHAPTER 2
BIRD OF DREAMS

1 Joan L. Watson, 'Notornis Rediviva', in *The Takahe: Fifty Years of Conservation Research*, eds William Lee and Ian Jamieson (Dunedin: University of Otago Press, 2001), pp. 23–30.

2 Dr G. B. Orbell, 'In Search of the "Extinct" Takahe – The Story of Its Rediscovery and Capture', *London Illustrated News*, 1 January 1949. All other quotes from Geoffrey Orbell in this chapter are from the same source.

3 Joan Watson, interview by Alison Ballance, 'Takahē – Back from the Brink', *Our Changing World*, Radio New Zealand, 29 September 2016, https://www.rnz.co.nz/national/programmes/ourchangingworld/audio/201817903/takahe-back-from-the-brink

4 "Gained in Telling: Recent Discovery of Notornis – An American Article," *Otago Daily Times*, 8 January 1949.

5 John W. Fitzpatrick, foreword to *The Takahe: Fifty Years of Conservation Research*, eds William Lee and Ian Jamieson (Dunedin: University of Otago Press, 2001), p. 9.

CHAPTER THREE
TAONGA

1 Kelly Keane-Tuala, 'Ngā Manu – Birds – Birds' Names', Te Ara – the Encyclopedia of New Zealand, accessed 29 July 2022, http://www.TeAra.govt.nz/en/nga-manu-birds/page-6

2 'Maukatakahea', Kā Huru Manu, accessed 28 July 2022, https://www.kahurumanu.co.nz/atlas

3 Estelle Pera-Leask, in discussion with the author, 29 June 2022. All quotes from Estelle in this chapter are taken from this interview.

4 The information here on place names comes from Kā Huru Manu (https://www.kahurumanu.co.nz/atlas) and the New Zealand Gazetteer (https://gazetteer.linz.govt.nz).

5 The information here on place names comes from 'The Takahē Story' signboard at Punanga Manu o Te Anau / Te Anau Bird Sanctuary.

CHAPTER FOUR
GIVE A BIRD A NAME

1 Gideon Mantell's address to the Zoological Society of London, 12 November 1850, quoted by Walter Buller, 'Notornis Mantelli (Mantell's Notornis)', *A History of the Birds of New Zealand* (London: Sir Walter Lawry Buller, 1888), p. 87, accessed 24 August 2022, https://nzetc.victoria.ac.nz/tm/scholarly/tei-BulBird-t1-g1-t2-body-d36.html

2 John Hall-Jones, *Fiordland Explored: An Illustrated History* (Wellington: A.H. & A.W. Reed, 1976), p. 29.

3 Brian Reid, 'Sightings and Records of the Takahe (*Notornis mantelli*) Prior to its 'Official Rediscovery" by Dr G.B. Orbell in 1948', *Notornis* 20, no. 4 (December 1974): 282.

4 Susanne Hill, John Hill and Victoria Jaenecke, eds, *Letters of a Naturalist – The Field Accounts of Richard Henry of Resolution Island* (Lower Hutt: Putangi Publications, 2022).

5 Graham Langton, ed., *Mr Explorer Douglas – John Pascoe's New Zealand Classic* (Christchurch: Canterbury University Press, 2000), pp. 224–225.

6 Ibid, p. 225.

7 Richard Henry to James. P. Maitland, 30 December 1898, quoted in Hill et al., *Letters of a Naturalist*.

8 Ibid.

9 Richard Henry to James P. Maitland, 17 January 1899, quoted in Hill et al., *Letters of a Naturalist*.

10 Richard Henry to J. P. Maitland, 7 June 1898, quoted in Hill et al., *Letters of a Naturalist*.

11 Robert Falla, '*Notornis* Re-discovered', *Emu – Austral Ornithology* 48, no. 4 (1949): 316–322, doi: 10.1071/MU948316

CHAPTER FIVE
FROM DREAM TO REALITY

1 Dr G. B. Orbell, 'Notornis Valley', *Illustrated London News*, 26 February 1949, p. 273. All other quotes by Orbell in this chapter are from this source.

2 Graham Turbott, 'Winter Observations on Notornis in 1949', *Notornis* 4, no. 5 (July 1951): 107–113.

CHAPTER SIX
THE HARD WORK BEGINS

1 R. H. Wheeler, 'Expedition into Fiordland', *New Zealand Geographer* 11, no. 2 (October 1955): 173–182.

2 Roger Duff, 'Recent Maori Occupation of Notornis Valley, Te Anau', *Journal of the Polynesian Society* 61, nos 1–2 (1952): 90–119, https://www.jps.auckland.ac.nz/document/?wid=2623

3 Brian Reid, 'Some Features of Recent Research on the Takahe (*Notornis mantelli*)', *Proceedings (New Zealand Ecological Society)* 14 (1967): 79–87, https://www.jstor.org/stable/24061281

4 Brian Reid, 'Sightings and Records of the Takahe (*Notornis mantelli*) Prior to Its "Official Rediscovery" by Dr G. B. Orbell in 1948', *Notornis* 21, no. 4 (December 1974): 277–295.

CHAPTER SEVEN
A CLOAK OF SECRECY

1 'Reared in Captivity for the First Time: Chicks of the "Extinct" New Zealand Takahe', *Illustrated London News*, 4 April 1959, p. 576.
2 Brian Reid, 'Takahe at Mt Bruce', in *Seminar on the Takahe and its Habitat 1978* (Invercargill: Fiordland National Park Board, 1978), pp. 96–117.
3 Brian Reid, in discussion with the author, 17 July 2019.

CHAPTER EIGHT
DEER YEARS

1 John Parkes et al., 'The Establishment and Control of Deer in the Murchison Mountains and Trends in Their Distribution and Density', in *Seminar on the Takahe and its Habitat 1978*, (Invercargill: Fiordland National Park Board, 1978), pp. 155–172.
2 Ibid.
3 Arthur Harper, letter to Ornithological Society of New Zealand, 23 April 1951, published as 'Protection of Takahe', *Notornis* 4, no. 5 (July 1951): 118.
4 Gordon Williams, 'Notornis in March, 1951', *Notornis* 4, no. 8 (April 1952): 202–208.
5 Michael Kelly, *Biodiversity Huts, Fiordland: Historic Heritage Assessmen*t (Wellington: Department of Conservation, 2010), p. 16.
6 Ibid, p. 28.

CHAPTER TEN
HOW MANY TAKAHĒ?

1 Brian Reid, 'Some Features of Recent Research on the Takahe (*Notornis mantelli*)', *Proceedings (New Zealand Ecological Society)* 14 (1967): 79–87, *https://www.jstor.org/stable/24061281*

CHAPTER ELEVEN
BIOLOGISTS AT WORK

1 Dr Jim Mills, in discussion with the author, 28 January 2019.
2 Professor Alan Mark, in discussion with the author, 30 September 2019.

CHAPTER TWELVE
MANAGEMENT IN THE MURCHISONS

1 Andy Garrick, pers. comm., 19 December 2019.
2 New Zealand Parliamentary Debates (Hansard) 468 (12 December 1985): 9178.

CHAPTER THIRTEEN
BURWOOD BEGINNINGS

1 Daryl Eason, in discussion with the author, 16 December 2019.

2 Peter Moore, pers. comm., 25 September 2021.

CHAPTER FIFTEEN
TO THE STUART MOUNTAINS

1 Glen Greaves, in discussion with the author, 27 August 2021.

2 Alan Tennyson and Paul Martinson, *Extinct Birds of New Zealand* (Wellington: Te Papa Press, 2006), p. 68.

CHAPTER SIXTEEN
ANATOMY OF A MAST YEAR

1 Roger Lavers, in discussion with the author, January 2019; Jim Mills, in discussion with the author, January 2019.

CHAPTER SEVENTEEN
STOAT THREAT

1 Jim Mills and Roger Lavers, pers. comm., 14 September 2021.

2 Ibid.

3 Robert Falla, 'The Nesting Season of Notornis', *Notornis* 4, no. 5 (July 1951): 97–100.

4 Jane Maxwell, 'Fiordland Takahē: Population Trends, Dynamics and Problems', in *The Takahē: Fifty Years of Conservation Management and Research*, eds William Lee and Ian Jamieson (Dunedin: University of Otago Press, 2001), pp. 61–79.

CHAPTER TWENTY-TWO
THE 2007 STOAT PLAGUE

1 Takahē Recovery Programme, 'Annual Report 2004–2005, unpublished Department of Conservation report.

2 David Steadman, 'Prehistoric Extinctions of Pacific Island Birds: Biodiversity Meets Zooarchaeology', *Science* 267, no. 5201 (24 February 1995): 1123–1131.

3 Ibid.

CHAPTER TWENTY-FOUR
A CHANGE OF DIRECTION

1 Glen Greaves, pers. comm., 2021. All quotes by Glen in this box are from the same source.

CHAPTER TWENTY-FIVE

GENETIC MANAGEMENT

1 Glen Greaves, in discussion with the author, 27 August 2021.

2 Department of Conservation, 'Takahē Recovery Goals 2021–2026', unpublished report.

CHAPTER TWENTY-NINE

THE RETURN

1 Phil Marsh, pers. comm., 11 May 2020. All other quotes by Phil in this box are from this source.

CHAPTER THIRTY

KAHURANGI – A NEW WILD HOME

1 Glen Greaves, interview by Alison Ballance, 'Turnaround in Takahē's Fortunes', *Our Changing World*, Radio New Zealand, 15 March, 2018, https://www.rnz.co.nz/national/programmes/ourchangingworld/audio/2018635907/turnaround-in-takahe-s-fortunes

2 Glen Greaves, pers. comm., 2018.

CHAPTER THIRTY-THREE

THINK BIG

1 Glen Greaves, in discussion with the author, 27 August 2021.

SPECIES LIST

COMMON NAME(S)	SCIENTIFIC NAME(S)
Auckland Island rail	*Porphyrio madagascariensis*
Auckland Island rail	*Lewinia muelleri*
Australian coot	*Fulica atra*
	Aciphylla takahea
banded rail, moho pererū	*Gallirallus philippensis*
bellbird, korimako	*Anthornis melanura*
black robin, karure	*Petroica traversi*
black swan, kakīānau	*Cygnus atratus*
blue tussock	*Poa colensoi*
brown creeper, pīpipi	*Mohoua novaeseelandiae*
Chatham Island coot	*Fulica chathamensis*
Chatham Island rail, mātirakahu, Hutton's rail	*Cabalus modestus*
cocksfoot	*Dactylis glomerata*
common brushtail possum, possum	*Trichosurus vulpecula*
common pheasant	*Phasianus colchicus*
copper tussock[1]	*Chionochloa rubra* subsp. *cuprea*
crown fern, petipeti	*Lomaria discolor*
curled snow tussock	*Chionochloa crassiuscula* subsp. *torta*
Dieffenbach's rail, moeriki	*Gallirallus dieffenbachii*
Eurasian skylark, kairaka	*Alauda arvensis*
ferret	*Mustela furo*
Fiordland tokoeka, subspecies of southern brown kiwi	*Apteryx australis* subsp.

1 Early scientific literature on takahē and deer diet refers to red tussock (Chionochloa rubra), but botanists now recognise that what they used to call red tussock in Fiordland is actually the subspecies copper tussock.

COMMON NAME(S)	SCIENTIFIC NAME(S)
great spotted kiwi, roroa	*Apteryx maxima*
grey teal, tētē-moroiti	*Anas gracilis*
Haast's carrot	*Anisotome haastii*
Haast's eagle, harpagornis	*Aquila moorei*
Hawaiian goose, nēnē	*Branta sandvicensis*
Hawkins' rail, mehonui	*Diaphorapteryx hawkinsi*
hihi, stitchbird	*Notiomystis cincta*
Hodgens' waterhen	*Tribonyx hodgenorum*
house sparrow, tiu	*Passer domesticus*
hunangamoho, broad-leaved bush tussock	*Chionochloa conspicua*
kagu	*Rhynochetes jubatus*
kākā	*Nestor meridionalis*
kākāpō	*Strigops habroptilus*
kakī, black stilt	*Himantopus novaezelandiae*
kea	*Nestor notabilis*
kurī, Polynesian dog	*Canis familiaris*
Macquarie rail	*Hypotaenidia philippensis macquariensis*
Madagascar teal, Bernier's teal	*Anas bernieri*
marsh crake, kotoreke	*Zapornia pusilla*
mid-ribbed snow tussock	*Chionochloa pallens* subsp. *cadens*
miro	*Pectinopitys ferruginea*
mohua, yellowhead	*Mohoua ochrocephala*
mountain beech	*Fuscospora cliffortioides*
mouse	*Mus musculus*
Mount Cook buttercup, giant buttercup	*Ranunculus lyalli*

COMMON NAME(S)	SCIENTIFIC NAME(S)
narrow-leaved snow tussock[2]	*Chionochloa rigida* subsp. *amara*
New Zealand coot	*Fulica prisca*
New Zealand falcon, kārearea	*Falco novaeseelandiae*
New Zealand fantail, pīwakawaka	*Rhipidura fuliginosa*
New Zealand pipit, pīhoihoi	*Anthus novaeseelandiae*
North Island adzebill	*Aptornis otidiformis*
North Island giant moa, kuranui	*Dinornis novaezealandiae*
North Island kōkako, kōkako	*Callaeas wilsoni*
North Island saddleback, tīeke	*Philesturnus rufusater*
North Island takahē, moho	*Porphyrio mantelli*
Norway rat	*Rattus norvegicus*
orange-fronted parakeet, kākāriki karaka	*Cyanoramphus malherbi*
Pacific rat, kiore	*Rattus exulans*
paradise shelduck, pūtangitangi	*Tadorna variegata*
perch	*Perca fluviatilis*
Petrie's mountain daisy	*Celmisia petriei*
pūkeko, Australasian swamphen	*Porphyrio melanotus melanotus*
rautahi, cutty grass	*Carex coriacea*
red beech	*Fuscospora fusca*
red deer	*Cervus elephus*
red tussock	*Chionochloa rubra* (including subspp. *cuprea*, *rubra* and *occulta*)[3]
rimu	*Dacrydium cupressinum*
rock wren, tuke, pīwauwau	*Xenicus gilviventrus*

2 The scientific literature on takahē and the Murchison Mountains often refers to broad-leaved snow tussock (*Chionochloa flavescens*). However, scientists have now realised that this species doesn't occur in the region, and the one that does is actually the narrow-leaved snow tussock (*C. rigida* subsp. *amara*).

3 The subspecies of red tussock that occurs in the Murchison Mountains is commonly known as the copper tussock (*Chionochloa rubra* subsp. *cuprea*).

COMMON NAME(S)	SCIENTIFIC NAME(S)
round-leaved snow tussock, terete-leaved snow tussock	*Chionochloa teretifolia*
ship rat	*Rattus rattus*
shore plover, tuturuatu	*Thinornis novaeseelandiae*
silver beech	*Lophozonia menziesii*
snipe-rail	*Capellirallus karamu*
southern brown kiwi, tokoeka	*Apteryx australis*
South Island adzebill, ngutu hahau	*Aptornis defossor*
spotless crake, pūweto	*Zapornia tabuensis*
stoat	*Mustela erminea*
takahē, South Island takahē	*Porphyrio hochstetteri*
thousand-leaved fern	*Hypolepis millefolium*
tomtit, miromiro	*Petroica macrocephala*
tuatara	*Sphenodon punctatus*
wapiti, elk	*Cervus canadensis*
weasel	*Mustela nivalis*
weka	*Gallirallus australis*
whio, blue duck	*Hymenolaimus malacorhynchos*
white heron, kōtuku	*Ardea alba*
yellow-crowned parakeet, kākāriki	*Cyanoramphus auriceps*

Bird names generally follow New Zealand Birds Online https://nzbirdsonline.org.nz
Plant names generally follow the New Zealand Plant Conservation Network https://www.nzpcn.org.nz

BIBLIOGRAPHY

I have used a wide variety of reference material, both published and unpublished. Unpublished material includes annual reports on the Takahē Recovery Programme. Many of the early scientific publications on takahē are published in the Birds New Zealand journal *Notornis*, which is searchable online.

Ballance, A. 2018. *Kākāpō: rescued from the brink of extinction.* Potton and Burton, Nelson.

Fiordland National Park Board. 1978. *Seminar on the takahē and its habitat – proceedings.* Fiordland National Park Board, Invercargill.

Hall-Jones, J. 1976. *Fiordland Explored: an illustrated history.* A.H. and A.W. Reed Ltd, Wellington.

Hill, S., Hill, J. 2015. *Richard Henry of Resolution Island: a biography.* Cadsonbury Publications, Christchurch.

Hill, S., Hill, J., Jaenecke, V. eds. 2022 *Letters of a Naturalist – The Field Accounts of Richard Henry of Resolution Island.* Lower Hutt, Putangi Publications.

Langton, G. ed. 2000. *Mr Explorer Douglas: John Pascoe's New Zealand Classic.* Canterbury University Press, Christchurch.

Lee, W., Jamieson, I. eds. 2001. *The Takahē: fifty years of conservation management and research.* University of Otago Press, Dunedin.

Mark, A. 2021. *Above the Treeline: a nature guide to alpine New Zealand.* Potton and Burton, Nelson.

Morris, M, Smith H. 1989. *Wild South: saving New Zealand's endangered birds.* Century Hutchinson NZ Ltd, Auckland.

Tennyson, A., Martinson, P. 2006. *Extinct Birds of New Zealand.* Te Papa Press, Wellington.

INDEX

Page numbers in **bold** refer to images.

C

D

E

H

I

J

K

L

M

N

O

P

Q

R

S

U

V

W

Y

Z

Published in 2023 by Potton & Burton
319a Hardy St, PO Box 221,
Nelson, New Zealand
www.pottonandburton.co.nz

© Alison Ballance
© Photographs: individual photographers

Edited by Susi Bailey

ISBN 978 1 98 855047 3

Printed in China by
Midas Printing International Ltd

This book is copyright. Apart from any fair dealing for the purposes of private study, research, criticism or review, as permitted under the Copyright Act, no part may be reproduced by any process without the permission of the publisher.

Detail of takahē feathers at beginning of Part 1, 2, 3 and 4. CRAIG MCKENZIE

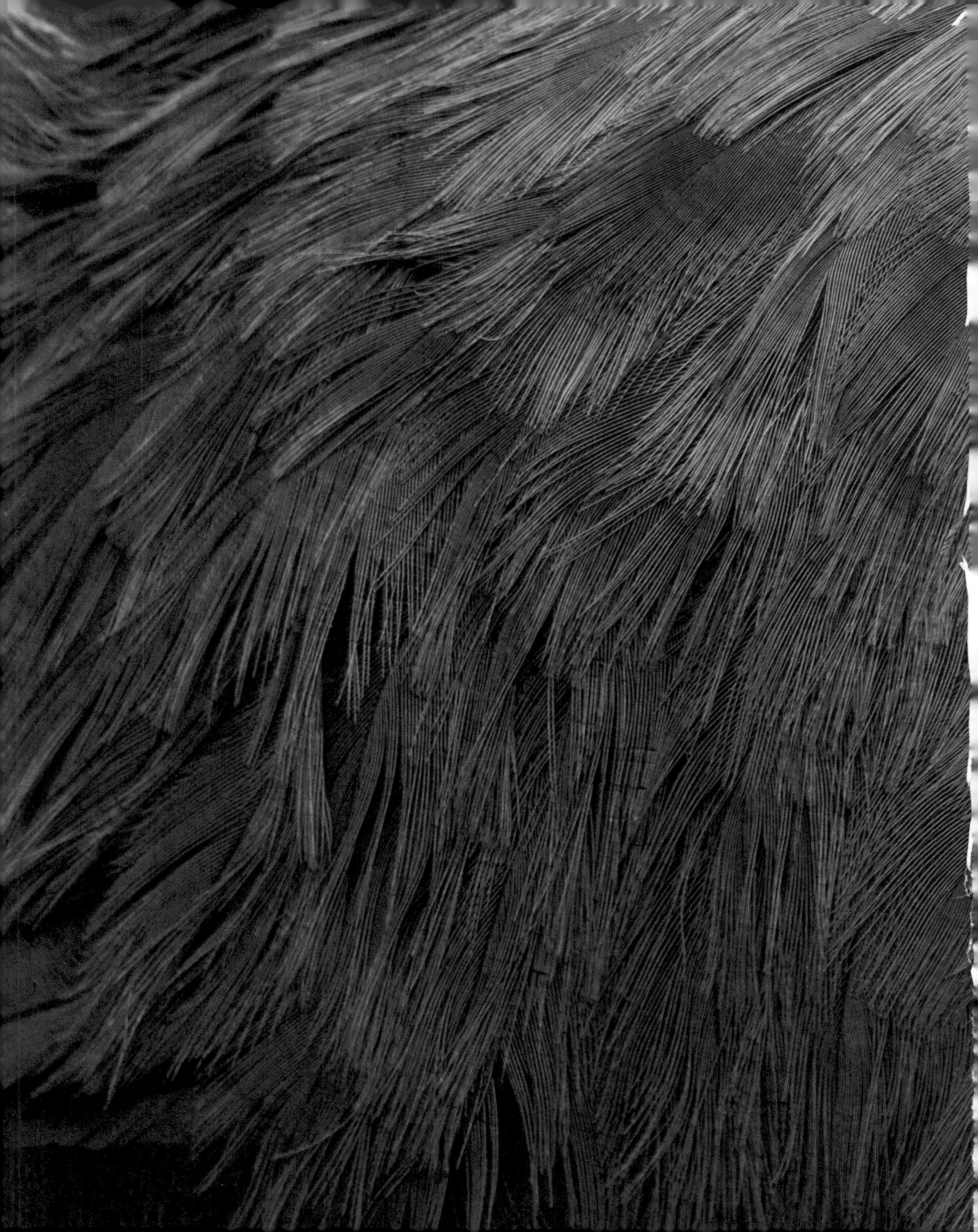